Weather and Climate
of the Great Lakes Region

Weather and
of the Great

UNIVERSITY OF
NOTRE DAME PRESS

Science
Technology
Health

Climate
Lakes Region

Val L. Eichenlaub

CARTOGRAPHY BY
Thomas W. Hodler

Library of Congress Cataloging in Publication Data

Eichenlaub, Val. L., 1933–
  Weather and climate of the Great Lakes region.

  Includes bibliographical references and index.
  1. Great Lakes region—Climate. I. Title.
QC982.8.E3   551.6′9′77   78–51526
ISBN 0–268–01929–0
ISBN 0–268–01930–4 pbk.

Manufactured in the United States of America

For my parents

Science
Technology
Health

Science
Technology
Health

Contents

On the weather of the Straits of Mackinac, where several of the Great Lakes come together:

> *This spot is midway between three great Lakes which surround it and seem to be incessantly playing ball with one another—the winds from the Lake of the Illinois no sooner subsiding than the Lake of the Hurons sends back those which it has received, whereupon Lake Superior adds others of its own. Thus they continue in endless succession; and, as these Lakes are large, it is inevitable that the winds arising from them should be violent, especially through the autumn.*

—Fr. Marquette

But in the days of good Father Marquette, Michilimackinac was indeed an earthly paradise. The sky hereabout was unusually clear; light breezes, wafting over the wide waters, brought relief in the warmest days; the air was freighted with the odor of the balsam.

—Reuben Gold Thwaites, *Father Marquette*

Preface

AT TIMES I HAVE SOME DOUBTS ABOUT GREAT LAKES weather. I suspect conspiracies of unknown origin enmeshing area residents in almost unimaginable weather. I felt that way after the winter of 1976–77 and for weeks following the blizzard of 1978. These are momentary lapses. A more pervasive emotion is a feeling of familiarity and expectancy, stirred by many seasons of Great Lakes weather. The first arctic outbreak of the late fall; the winter periods of snow squalls alternating with brilliant sunshine; the slow return of warmth in the spring; the bursts of heat and humidity which mar every summer—these instill a sense of yearly regimen.

Recently, I spent several weeks in southern California. As it was then the dead of winter in Michigan, I expected the weather of California to offer better things. And it did, to a degree. But it also offered the unfamiliar. The Pacific Ocean seemed capable of generating all sorts of strange weather patterns. By satellite, they appeared as vast globs of cloud spawned in the tropics or as Aleutian storms with long, comma-shaped appendages of cloud sweeping southwestward. It was difficult to tell just what these storms were doing, or were capable of doing, until they arrived (if they did). The weather didn't behave at all like Great Lakes weather, and I suffered from a constant feeling of disorientation.

Conspicuously missing from the California weather mélange were the lake effects, so prominent during Michigan winters. The lake effects are part of and contribute to the familiar touch that I associate with Great Lakes weather, and are, in fact, what this book is all about. It seemed not mere chance that on the return flight between O'Hare International Airport in Chicago and Kalamazoo, I encountered the telltale increase of cloud over Lake Michigan and arrived home in a blinding lake-effect snowstorm.

I will probably continue to have mixed reactions to Great Lakes weather: berating it on occasion, but basking in the familiarity of its yearly cycles. Writing a book about it has been an exciting and challenging task, one which could not have been accomplished without the assistance and cooperation of a number of people. I have also had available the studies of many competent researchers who have devoted their careers to understanding Great Lakes weather.

It is impossible to name all those who have helped in one way or another, but special thanks should go to the following people: Walter Lyons, for his wide-ranging research on mesocale atmospheric processes occurring over and near the Great Lakes, and for the use of photographs from his extensive collection; Jay Harman, Ken Dewey, and Margaret Lemone for supplying photographs of various Great Lakes weather features; R. C. Peterson and Helen Johnson for providing photographs of the Keweenaw Peninsula of Michigan; Stanley A. Changnon, Jr., and the staff of the Atmospheric Sciences Section of the Illinois State Water Survey for their countless climatological studies of the Great Lakes; Thomas Hodler for the design and drafting of diagrams; Paul Krause for assistance in assembling the climatic records; and the department of Geography of Western Michigan University, for funding portions of the research.

I am also indebted to Kayla Wilson and my mother, Mrs. W. V. Eichenlaub, for typing early drafts of the manuscript, and to my wife, Chris, who critically examined early drafts of several chapters.

Mark O'Connell was the production editor, and I am grateful for his patience in overseeing the numerous production details.

The assistance of each person is greatly appreciated. Any errors of fact or interpretation are, of course, my own.

Weather and Climate
of the Great Lakes Region

Introduction

THE WINTER OF 1976–77 BEGAN EARLY AND RELENT-lessly battered the United States for 4 months.

The East was buffeted by blizzard after blizzard and experienced, according to the National Weather Service, its coldest winter since the founding of the Republic. Fuel bills skyrocketed, natural gas shortages caused closings of schools and factories, industrial production declined, roads were blocked by huge snowdrifts, and rivers were clogged with ice. Snow fell for the first time ever in Miami, and losses to the Florida citrus industry mounted because of freezing temperatures. In Ohio, one million workers were laid off as a result of the natural gas shortage. In upstate New York, Buffalo and Watertown were hit by record snows. Nothing moved for days as besieged residents slowly dug their way out.

With average January temperatures of 25.1°F (−3.8°C) in the East setting an all-time record for cold (the previous January record was 25.3°F in 1857), the weather made headline news day after day. The weather also made headlines in the West —but it was a different story there, with record warmth and drought. Reservoirs ran dry, water supplies became critical, and the mountain slopes, where many feet of snow normally attracted skiers by the droves and provided an invaluable storage facility for much-needed irrigation water, lay bare in the sunshine. Numerous state-financed cloud-seeding projects were spawned by the unusually dry and warm winter.

The Midwest, including the Great Lakes, also experienced its coldest January of record. Temperatures averaged 11.3°F (−11.5°C), even lower than the previous record in 1857 of 12.9°F (−10.6°C). The Ohio and Mississippi rivers froze solid in stretches, and ice on the Great Lakes reached maximum extent. Record snows accompanied the extreme cold, taxing snow-removal crews and county and municipal budgets. Residents were urged to "dial down," and fuel supplies became short.

The worst blow of the winter was hurled at the Great Lakes on 26–31 January. A storm roared out of the northwest and exploded over Lake Michigan. Below-zero cold, driving snow, and 40-mile-per-hour winds paralyzed western Michigan for days. In Buffalo, New York, snow accumulations climbed over 40 inches (102 centimeters); at Watertown, New York, over 60 inches (152 centimeters). Metropolitan Buffalo was declared a disaster area, and economic losses were placed at $150 million. In southwestern Michigan, big "up-north" snow movers were brought in to free the clogged rural roads. Governor William Milliken asked President Carter to declare thirty Michigan counties as disaster areas. The damage toll from the January 26–31 blizzard rose to nearly $8 million. Officials in Van Buren County, Michigan, stated that nearly half of their yearly road and highway budget was already spent during the months of January and February!

Culpable for the unusual weather was the wind pattern aloft over North America. Instead of flowing from west to east, as is usually the case, the jet stream (the core of westerly winds which occur at the higher levels) was blocked along the Pacific Coast by an enormous and persistent high-pressure area. A deviant route far north toward Alaska, then southward over the eastern United States, was the result. It was late February before the jet was able to crack the blocking high, bringing much-needed moisture to the western states and allowing the East to be warmed by gentle breezes from the Gulf of Mexico. The winter of 1977–78 began innocuously enough. The Great Lakes region escaped the severe cold of the preceding winter, and the big storms were traveling far to the South and then turning up the eastern seaboard. But, on January 25, the surface weather chart appeared ominous. An intensifying storm over northern Minnesota was moving rapidly southeastward, and a developing low pressure system far to the South, over northern Louisiana, was drifting slowly toward the northeast.

During the night, the southern storm accelerated northward through eastern Kentucky and Ohio and met the northern storm over Lake Erie. Over much of the Great Lakes area, the result was one of the most severe blizzards on record. The central pressures of 28.23 inches of mercury were comparable to those recorded in the eye of a hurricane, and were the lowest on record for many Great Lakes cities. Heavy snows and driving winds paralyzed Michigan, Indiana, Ohio, and Ontario. Schools in some areas remained closed for almost two weeks.

Although January temperatures were less frigid than in 1977, the cold's persistence was remarkable. In some parts of the Great Lakes, February turned out to be one of the coldest months on record. Meanwhile, on the West Coast, a drought ended and storm after storm caused record rains, flooding, and mudslides in California.

These were winters to remember. The nation was sharply reminded that despite a mind-boggling age of technological opulence, the weather is still a factor to be reckoned with.

This had always been true for our forefathers. Decades ago, weather determined the daily run of activities much more than today. Lives were structured by constant, enervating encounters with the elements. Weather descriptions crowd the diaries and journals of the early explorers and settlers, reminding us of the acute awareness these people had of how weather and climate controlled their lives. Perceptions differed, as attested by conflicting statements of Father Marquette and Reuben Gold Thwaites—and what was good weather for some was bad weather for others, just as today. There can be little doubt, however, that weather, good or bad, has played an important role in the history of the Great Lakes.

We now seem to be reapproaching the vulnerability our ancestors exhibited to weather. Bolstered by decades of unlimited energy and burgeoning grain supplies, we had incurred an unwarranted disdain for the weather in the decades prior to the 1970s. But the severe winters of 1976–77 and 1977–78, along with other "unusual" weather events in the early and middle 1970s, combined with spotty food shortages and sharply increasing energy prices to create a renewed respect and interest.

Weather is now front-page news, and no individual is immune to its antics and vagaries. Decision making, from personal to international levels, has become increasingly responsive to weather, and there is no indication of decreased impact in the future. It is not surprising that efforts to understand and predict the weather have accelerated in recent years.

This is a book about the weather and climate of the Great Lakes of North America. It is about good, bad, normal, and abnormal weather, and about climates of the past, present, and (with some trepidation) the future. But it particularly concerns those weather features which uniquely belong to the Great Lakes themselves. What so often escapes us is that the Lakes themselves are weather factories capable of causing twists of climate found in few other parts of the world.

These are weather types which we, as natives, tend to take for granted. Arising chiefly from the presence of the Lakes, these types lend the area a character all its own. For the weather buff, they make the Great Lakes one of the most interesting regions in the world. And for the forecaster, they make it one of the most exciting.

The General Setting
of the Great Lakes

ALTHOUGH OCEAN-LIKE IN MANY WAYS, THE GREAT Lakes were actually formed from ice sheets nurtured by the vast continentality of North America. On the time scale utilized by the geologist, the Lakes are very recent features, only 10,000–12,000 years old. They are constant reminders of a past cold age and of the tons of ice which rested over the area during the *Pleistocene,* the epoch of geologic history which has occupied the past several million years. This period was characterized by a number of glacial advances southward into the United States.

The immense accumulation of ice—thousands of feet in thickness—gouged, scraped, and polished the ancient, resistant rocks of Canada; but it more easily enlarged the glacial troughs composed of softer rocks to the south. These troughs were eventually to contain the Great Lakes. As the glaciers receded, the melting waters accumulated within these basins to form the predecessors of the Lakes which we now know. The present outlines of the Lakes are the result of an evolutionary period lasting thousands of years, during which the ancestral lake levels rose and fell, their shorelines receded and advanced, and their drainage outlets changed several times.

Various channels emptied the surplus impounded waters.

First, the waters found their way to the sea through a channel near the site of present-day Chicago, thence down the Mississippi drainage system to the Gulf. Later, the Hudson-Mohawk valley route in present-day New York State provided an outlet to the Atlantic. Eventually, the present outlet via the St. Lawrence River to the Atlantic was formed.

Throughout the area, ample and convincing evidence points to the varying levels and configurations of these ancestral lakes. Spillways and pondings, wave-cut shore bluffs, ancient beach ridges, and varved clays are indicative of a time when the Lakes contained more water than they now do and would have presented unfamiliar outlines on a map. There was also a time when the Lakes contained less water than they presently do. At that time, the surface waters of Lake Michigan were almost 300 feet below the present level, and the areal extent of the lake was reduced to what might be described as a mere puddle of water.

Centuries elapsed while the glaciers were retreating into Canada and the Lakes were organizing and reorganizing their configurations into their present patterns. During these centuries, a climate existed which was significantly different from the one we now experience. It was generally colder, reminiscent of the ice sheets which had formerly buried the land. As the climate moderated with removal of the ice into far northern Canada, vegetation slowly appeared—first, the plant associations common to the boreal forest regions of present-day Canada, but later the associations which, had man not so profoundly altered the landscape during recent centuries, would be familiar to us today.

Locational Lineaments of the Great Lakes

Of major importance to the present climate of the Great Lakes is the fact that they were formed roughly halfway between the equator and the North Pole within a lowland corridor. This corridor extends, without interruption by mountains, through eastern North America from the shores of the

Arctic Ocean to the Gulf of Mexico. A glance at the map will show that the forty-fifth parallel of latitude, which is halfway between the equator and the Pole, almost equally divides the waters of the Lakes; and that the North American continent is wedge-shaped and broadens toward the polar latitudes. The open corridor in the eastern portion of the wedge allows an easy exchange of frigid Arctic air from the north and balmy subtropical air from the south and is duplicated nowhere else in the world. The fact that the glaciers formed the Lakes midway within this corridor is of the utmost meteorological importance.

Six in number, the Great Lakes of North America extend through a latitudinal span of over 7½ degrees. The most northerly indentation of Lake Superior, Nipigon Bay (figure 1), is located at a latitude of 48°58′ N. The shoreline of Lake Erie, nearly 575 miles away, bends southward west of Cleveland to a latitude of 41°20′ N. For comparison, the latitudinal extent of the Lakes is about the same as that between the cities of Cleveland, Ohio, and Atlanta, Georgia—a significant span climatically.

The east-west extent of the Lakes is even greater. Duluth, Minnesota, located at the western end of Lake Superior, has a longitude of 92°07′ W. More than 800 miles to the east, at a longitude of 76°07′ W where the Great Lakes system empties into the St. Lawrence River, are located the Thousand Islands, remnants of a resistant sill of ancient rocks which extends from Canada into New York State. This is equivalent to the distance from Detroit to Jacksonville, Florida, or from Chicago to Dallas, Texas. When the drainage basin of the Lakes is included, the latitudinal and longitudinal dimensions of the system become even larger.

Configurations and Dimensions of the Great Lakes

The map indicates that the general configurations of the Lakes, as well as their individual latitudinal and longitudinal positions, are different. Lake Erie and Lake Ontario are

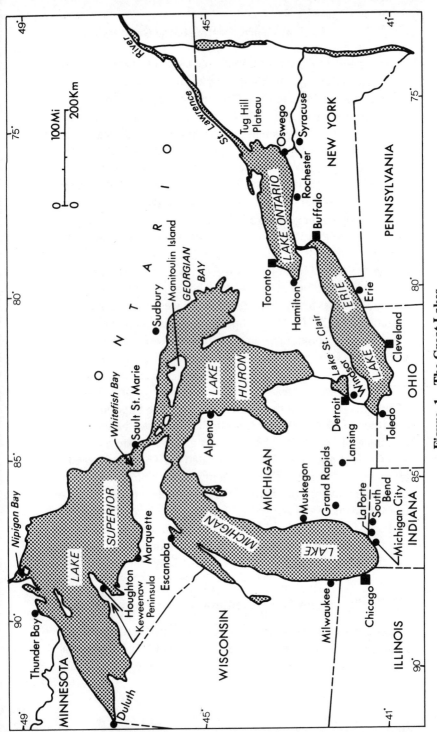

Figure 1 The Great Lakes

positioned in more southerly latitudes and are aligned essentially in an east-west direction. Lake Huron and Lake Michigan are aligned in a north-south direction and span a considerable latitudinal extent. Lake Superior is longer along its east-west axis, although its elongation is not as pronounced as that of the other Lakes. It is essentially located within higher latitudes. Lake St. Clair, of course, is very small; although considered a Great Lake, it lacks the impressive dimensions of the others.

Vital statistics for the Great Lakes are given in table 1. Water surface and drainage basin areas of each Great Lake, with the exception of Lake Michigan, are shared by the United States and Canada. The water surface area of the Great Lakes comprises more than one-third of their land drainage basin.

Lake Superior is the largest, deepest, and most northerly of the Great Lakes. The combined waters of all the other Great Lakes could easily be held within the basin which contains Lake Superior, and its areal extent is about equal to that of Lake Michigan and Lake Erie combined. Lake Huron, however, has the longest shoreline, with the indented, uneven shore of Georgian Bay, and its many islands including a very large one (Manitoulin).

Figure 2 shows a representative depth profile of the Great Lakes. All of the Lakes except Lake Erie and Lake St. Clair have bottom contours which extend below sea level, although only Lake Ontario has an *average* depth below sea level. Despite this fact, the Lakes contain fresh water.

Changes in the Level of the Lakes

The levels of the Lakes have varied on time scales ranging from thousands of years to hours. Variations of the ancestral lake levels as the glaciers receded have already been mentioned. Variations of present lake levels over indeterminate periods ranging from 5 to 20 years also occur. These variations are related to meteorological factors, primarily precipitation over

Table 1
VITAL STATISTICS OF THE GREAT LAKES

	Lake Superior	Lake Michigan	Lake Huron	Lake St. Clair	Lake Erie	Lake Ontario	All Lake
Water surface area in square miles	31,700 (82,103km^2)	22,300 (57,757km^2)	23,050 (59,670km^2)	490 (1,269km^2)	9,910 (25,667km^2)	7,550 (19,554km^2)	95,000 (246,050km^2)
Drainage basin land, in square miles	49,300 (127,687km^2)	45,600 (118,104km^2)	51,700 (133,903km^2)	6,930 (17,948km^2)	22,700 (58,793km^2)	27,200 (70,448km^2)	203,430 (526,883km^2)
Drainage basin land and water, in square miles	81,000 (209,790km^2)	67,900 (175,861km^2)	74,800 (193,732km^2)	7,420 (19,218km^2)	32,600 (84,434km^2)	34,800 (90,132km^2)	298,520 (773,166km^2)
Length of coastline including islands, in miles	2,980 (4,798km)	1,660 (2,673km)	3,180 (5,120km)	169 (272km)	856 (1,378km)	726 (1,669km)	9,571 (15,409km)
Maximum depth, in feet	1,333 (407m)	923 (282m)	750 (229m)	21 (6m)	210 (64m)	802 (245m)	—
Average depth, in feet	489 (149m)	279 (85m)	195 (59m)	10 (3m)	62 (19m)	283 (86m)	—
Volume of water, in cubic miles	2,935 (12,239km^3)	1,180 (4,921km^3)	849 (3,540km^3)	1 (4.17km^3)	116 (484km^3)	393 (1,639km^3)	5,474 (22,826km^3)
Mean elevation above sea level, in feet	600.37 (183m)	578.68 (176.4m)	578.68 (176.4m)	573.01 (174.7m)	570.37 (173.9m)	244.77 (74.6m)	—

ADAPTED FROM National Oceanic and Atmospheric Administration, Lake Survey Center, *National Ocean Survey*, HO 26-511 (11-71).

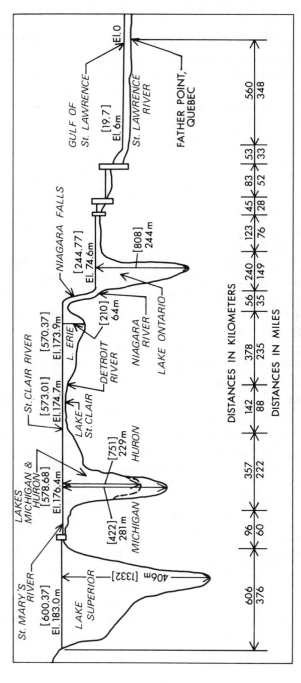

Figure 2 Profile of the Great Lakes–Saint Lawrence River System (bracketed measurements in feet)

the basin, and secondarily to evaporation rates. Over a period of 120 years, lake levels have varied as much as 6½ feet on Lake Ontario, 5½ feet on Lake Erie, and 4 feet on Lake Superior. During the early 1970s, the levels of the Great Lakes were high and much concern was expressed over the erosion of shorelines and the removal of beaches and lakefront cottages by storm-driven waves. The levels are now receding; in the fall of 1977, Lakes Superior, Michigan, and Ontario had dropped below the long-term mean. During the late 1950s and early 1960s, lake levels were very low, and attention focused on the maintenance of channels for lake shipping and the continued availability of a fresh-water supply.

A seasonal variation of lake levels also occurs. This variation is more predictable than the longer term oscillations just described. Lake levels are generally highest in the spring and summer, lowest in fall and winter. The seasonal change ranges from a little less than 2 feet on Lake Ontario to a bit more than 1 foot on Lakes Michigan and Huron. Peak levels during the spring and summer are related to the release of winter moisture which, over a large part of the basin, has been stored in the form of snow. In addition, precipitation amounts in the spring and summer are slightly higher than in the fall and winter. The low levels in winter are related to high evaporation rates, when lake waters are warmer than the air, and to the fact that runoff is minimal because much of the precipitation is stored in the form of snow to await the spring melting period.

Changes in lake levels also occur on a short-term basis, when atmospheric factors cause seiches or pileups of water, in certain portions of the lake basin. These will be discussed in more detail in a later chapter.

Mean lake levels, in feet and meters above sea level, are given in table 1. The level of Lake Superior is the highest, at approximately 600 feet (183 meters). Lake Superior's level can be controlled to some extent by regulation of outflow through the locks at Sault Ste. Marie. Increased or reduced outflow into the St. Lawrence River can also be employed to partially regulate the level of Lake Ontario. Both Lake Superior and Lake Michigan drain into Lake Huron—Lake Superior through the St. Mary's River and Lake Michigan through the Straits of Mackinac. As the mean level of Lake Huron is approximately 578 feet (176 meters) above sea level, the waters of Lake Superior drop about 22 feet (66 meters) on their journey southward through the St. Mary's River. The mean level of Lake Michigan is the same as that of Lake Huron.

The waters of Lake Huron empty into Lake St. Clair via the St. Clair River. A drop of about 6 feet (18 meters) gives Lake St. Clair an average level of 573 feet (175 meters). From Lake St. Clair, the Detroit River descends another 3 feet to Lake Erie's mean level of 570 feet (174 meters). The most spectacular difference in mean lake levels occurs between Lake Erie and Lake Ontario, which has a mean lake level of 244 feet (75 meters) above sea level. A large portion of this descent occurs at Niagara Falls, where the Niagara Cuesta, a ridgelike formation of dolomitic limestone, is gradually being eroded southward.

Implications for Weather and Climate

The meteorological implications of these statistics are considerable. Latitudinal differences of 8 degrees within the basin insure that significant contrasts of climate will exist between the northern and southern Lakes. The longitudinal extent of the Great Lakes system places the area within a zone of gradual transition from the continentality of the central North American continent to the more marine-like eastern borders. The varying alignments of the Lakes and their differences in areal extent combine to cause contrasting interactions of differing magnitudes that modify the climate of surrounding shores. And the varying depths of the Lakes influence the heating and cooling rates of the waters, their capacities to interact with overpassing air, and the probability of extensive ice formation during the winter. Changing lake levels enhance the probability that storms may damage shorelines during periods of high water and that sand deposited by storms and currents may block channels or necessitate additional dredging during low-level periods. There can be little doubt that the work of the glaciers thousands of years ago, by determining the present configurations and dimensions of the Lakes, has posed an important control of the present climate and weather.

PART ONE

The Atmospheric Controls

1: The Energy Source for Great Lakes Weather

ENERGY RESULTING FROM CONTINUOUS THERMO-nuclear reactions occurring under high temperatures deep in the sun's interior provides the basic fuel for all meteorological processes. This energy is also, directly or indirectly, the source of practically all energy utilization, with the exception of nuclear energy, by man.

Near the center of the sun, where temperatures are thought to be 15,000,000°C, hydrogen is converted to helium. This incessant production of energy is offset by energy radiated outward from the sun into space in the form of electromagnetic waves. This form of energy transfer requires no material medium. At the mean distance of the earth from the sun (about 93 million miles or 149 million kilometers), solar radiation is considerably attenuated—in fact the earth receives only about one five-billionth of the total energy given off from the sun into space. This amount, however, is sufficient to sustain all forms of life. It is the exchange of this energy within the earth-atmosphere system which is essentially responsible for most of the features of weather and climate which the Great Lakes experience.

The Nature of Radiation:
Solar and Terrestrial Radiation

All bodies radiate energy in the form of electromagnetic waves. Several basic laws of physics describe the nature and behavior of this radiant energy. The Stefan-Boltzman law states that the amount of energy emitted is a function of the temperature of the radiating body, with energy emission increasing as the fourth power of the temperature. More simply stated, the Stefan-Boltzman law states that a hot object gives off much more radiation per unit area than does a cold object.

The span of wavelengths in the electromagnetic spectrum is very large (figure 3), ranging from very longwave, low-frequency radio emissions (with wavelengths measured in kilometers) to very shortwave emissions such as X-rays and gamma rays (measured in angstrom units [one-millionth of a millimeter]). The Wien displacement law relates the wavelength of maximum emission inversely to the temperature of the radiating body. When we make a comparison of energy emitted from the sun and the earth, some striking contrasts become apparent. First, as the Stefan-Boltzman law tells us, any unit area on the sun's surface such as a square centimeter or kilometer emits an almost inconceivably greater amount of energy, one hundred sixty thousand times more than a similar area at the surface of the earth. Second, following the Wien displacement law, the wavelengths at peak emission of solar radiation are much shorter than those for the earth.

In fact, the sun, with a surface temperature around 6,000°C, radiates almost 50% of its energy within the visible wavelengths. The remainder is in wavelengths shorter than the visible span (ultraviolet), or as infrared radiation in the longer wavelengths. We cannot see radiation in these wavelengths. The earth, responding in a manner described by the Wien displacement law, radiates practically all of its energy in the longwave infrared segment as thermal or heat radiation invisible to the human eye.

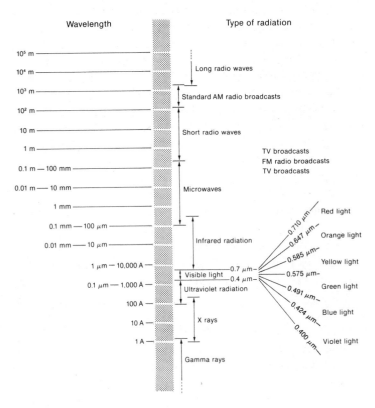

Figure 3   The Electromagnetic Spectrum

Variation of Solar Energy
Received at the Earth's Surface

A unit of measurement for solar energy is the *calorie*, which is defined as the amount of energy required to raise the temperature of 1 gram of water 1°C. Energy received per unit area is expressed in *langleys* (1 calorie per square centimeter). On an imaginary space platform at the top of the earth's atmosphere that directly faces the sun, the rate of energy reception (energy flux) from the sun is nearly constant. On our space platform, 1.95 langleys per minute are received at the mean distance of the earth from the sun. This value (radiant flux density) is known as the *solar constant*.

Of greatest concern to us is the amount of energy received at various places on the *surface* of the earth. The earth's surface is surrounded by an envelope of gases (the atmosphere), and surface areas are not usually oriented to receive the sun's direct rays. Energy reception at the earth's surface, therefore, is less than the solar constant, and the amounts vary in space and time. These variations of energy are the key to understanding many features of Great Lakes weather, and they form the driving mechanism for atmospheric processes.

It is essential that we briefly investigate why the energy amounts at the earth's surface vary with time and place and how this variation imposes an important control over Great Lakes weather and climate. It was stated that the solar constant value, 1.95 langleys, is the amount received at the top of the atmosphere on a surface normal (at right angles) to the incoming rays of the sun. This point is labeled *A* on figure 4. Examine for

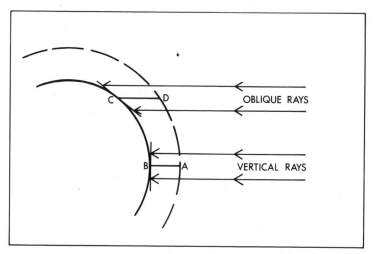

Figure 4 **Variations in the Amounts of Solar Energy Received by Different Points on the Earth's Surface**

a moment point B on the surface of the earth, directly beneath point A. Note that the angle at which the sun's rays are received remains the same, but that before this energy can reach point B it must pass through a portion of the atmosphere represented by the segment A-B. Now examine point C. Note that the sun's rays no longer shine from directly overhead but at an angle from the horizon, called the *angle of inclination* (solar altitude). Note also that the line C-D, representing the length of the path of the sun's rays through the atmosphere, is considerably longer than line A-B.

Amounts of solar energy received at point A are greater than those received at point B, while amounts at point B are greater than those received at point C. A close inspection of figure 4 should suggest some reasons why this is so. Let's try to answer the question, why are the amounts at point B less than those at point A? Note that at point A, which is outside the atmosphere, the energy does not pass through the atmosphere before being received. At point B, however, the energy has passed through an entire column of air, from top to bottom. The atmosphere is composed of a mixture of gases* and suspended particles. Very often it contains clouds. This composite of gases, particles, and clouds (consisting of tiny water droplets or ice crystals) has the capacity to deplete solar radiation in various ways.

Gases absorb radiation selectively, and some of the atmospheric gases can absorb rather well within the shorter wavelengths of solar radiation. Oxygen and ozone are two gases which absorb well in the very short (ultraviolet) wavelengths. Consequently, some of the energy from the sun is absorbed before reaching the earth's surface. But most atmospheric gases do *not* absorb well in the wavelengths primary to solar radiation; thus most of the solar radiation can pass through

*Dry air contains the following gases in approximate percentages by volume: nitrogen 78.08; oxygen 20.94; argon 0.93; carbon dioxide 0.03; and a host of gases in smaller quantities. Variables include water vapor 0–3%, and ozone 0.1–0. 2 parts per million at high levels, less at ground (Byers 1974, p. 31).

the atmosphere relatively unimpeded by absorption. However, dust particles and other impurities may scatter energy in certain wavelengths. Some of this energy is scattered back into space and never arrives at the surface. And if clouds are present (as they frequently are over large portions of the earth), larger amounts of solar energy may be reflected from cloud tops back into space and prevented from reaching the surface. Thus, because of absorption, scattering, and reflection, point *B* receives less energy than point *A*.

But (assuming that atmospheric conditions are about the same over both points) why does point *C* receive even less energy than point *B*? Notice that the sun's rays reach point *C* at an oblique angle. This fact has several serious implications for energy reception. First, the energy at point *C* is dispersed over a much larger area than at point *B*. The same effect can be produced by shining a flashlight beam on a dark table top. By shining the beam directly over the table top, the rays are concentrated within a relatively small surface area. But if the flashlight is beamed at an oblique angle to the table top, the rays are dispersed over a much larger area. Any unit area (such as a square centimeter) receives considerably less energy at the oblique angle than at the angle normal (a right angle) to the surface.*

Second, note that the rays of the sun must pass through a longer portion of the atmosphere before reaching the surface at point *C* than at point *B*. This means that more depletion can occur, as there is more atmosphere to be penetrated. The net result is that the amount of solar radiation received at point *C* is considerably less than that received at point *B*.†

At all points on the earth's surface, the angle at which the sun's rays are received plus the atmospheric capacity to deplete energy by absorption, scattering, and reflection are the factors

*Lambert's cosine law states that the radiation intensity varies as the cosine of the angle between the perpendicular and the angle of the incidence.

†Beer's law describes the reduction of intensity as a function of the depth of the atmosphere (path length) and an extinction coefficient.

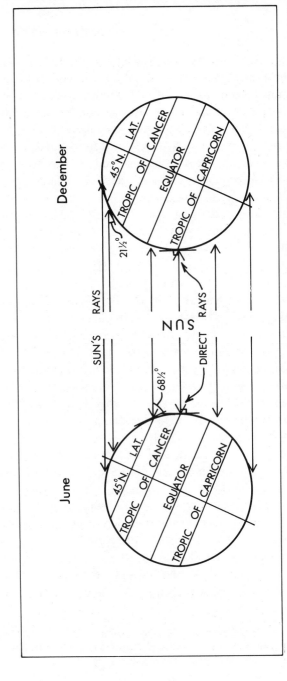

Figure 5 Contrasting Solar Angles at 45° N in June and December

which control the amount of solar radiation available to heat the surface and atmosphere. In addition, the variation in length of day must be considered; obviously, the longer the sun is above the horizon the greater the amount of energy which can be received. For any given point on the earth's surface, the angle of the sun's rays and the length of day can be predicted for any day and hour of the year. The condition of the atmosphere, however, cannot be predicted with any surety, and we can make only rough estimates of the amount of cloudiness or turbidity based upon past climatic averages.

Seasonal Variation of Solar Energy within the Great Lakes Region

To see how energy reception might vary seasonally in the Great Lakes area, let's make a simple computation of the angle of inclination, then determine the length of day at two times of the year: at summer solstice (about 21 June) and at winter solstice (about 21 December). As the angle of inclination varies according to the time of day (being minimum at sunrise and sunset and maximum at high noon, sun time), we will compute the angles at comparable times for solar noon. We will assume, for the purpose of this problem, that there is no difference in the transmissivity of the atmosphere.

The fact that the earth is tilted on its axis approximately 23½° from the perpendicular means that as it revolves around the sun, the Northern Hemisphere will be tilted toward the direct rays of the sun at one stage of its journey, while the Southern Hemisphere will receive the direct rays at another stage. Figure 5 shows the earth in relation to the rays of the sun at its summer and winter solstice positions. Note that at the summer solstice, the direct rays (90° angle of inclination) are received at the Tropic of Cancer, 23½° north latitude. At the latitude of the Great Lakes (45° N), the angle of inclination is large, but the sun's rays are not received from directly overhead. At the winter solstice, on the other hand, the direct rays of the sun reach the Tropic of Capricorn in the Southern Hemisphere, and the rays which reach the Great Lakes area are quite oblique. Figure 5 thus indicates that a large difference exists in the angle of inclination in the Great Lakes region between winter and summer. Anyone who has resided in the area is surely aware of this. In the summer, the sun at noon burns down from high above the southern horizon, and the fact that large amounts of energy are being received is apparent from the warmth of the sun. In the winter, the sun appears pale and wan, low on the southern horizon. The winter sun just doesn't seem capable of conveying much warmth.

The angular distance of the sun from the zenith at noon is always the difference between the given latitude (in this case the latitude of the Great Lakes, or 45° N) and the latitude where the sun is directly overhead. It follows, then, that on 21 December when the sun is directly overhead at 23½° south latitude, the noon sun shines from an angle of 68½° (45° plus 23½°) from the zenith, or 21½° from the southern horizon. On 21 June, however, the sun at high noon shines from an angle of 21½° from the zenith, or 68½° from the southern horizon. Thus a seasonal difference of 47° (the latitudinal difference between the Tropic of Cancer and the Tropic of Capricorn) exists in the angle of inclination.

In December, with the sun's rays at an oblique angle over the Great Lakes (21½°), the dispersion factor plus the fact that solar energy passes through a considerable thickness of the atmosphere before reaching us combine to reduce substantially the amount of solar energy received. Consider also that in December the duration of daylight averages about 8½ hours and that the amount of cloudiness is extremely large, and you can see why the Great Lakes area receives only very small solar radiation amounts in the winter—amounts averaging only about 105 langleys per day in December! During the summer when the angle of inclination is much greater, on the other hand, the days are longer and the cloud cover is minimal; about 530 langleys, or nearly five times the winter amount of energy, is received during an average day. As this solar energy heats the surface and ultimately the air, there can be little doubt as to why the Great Lakes region shows strong seasonal differences in temperature, experiencing relatively warm summers and quite cold winters. This large seasonal variation in energy reception is common to the middle latitude regions of the world and forms the basis for the marked change of seasons experienced by the temperate zones.

Disposition of Energy at the Earth's Surface

Energy which ultimately reaches the earth's surface can either be *absorbed,* thus heating the surface, or *reflected* back into the atmosphere. The *albedo* of a surface measures, in percent, the amount of energy reflected. Although the albedos of various earth surface materials vary, the largest differences exist between surfaces covered with snow or ice and those free of snow and ice.

Located in the middle latitudes, the Great Lakes region presents a surface covered with snow or ice for a number of months each year. Snow is a very good reflector (albedo of fresh snow ranges from 80% to 90%), and most of the energy received on a snow-covered surface is simply reflected back to the atmosphere. This is why, on sunny winter days with little wind, temperatures remain cold and snow does not melt, while people may feel fairly comfortable wearing even light clothing. The energy is being reflected from the snow with little absorption occurring; consequently, the snow does not melt and the air remains cold. Fairly large amounts of energy, however, are being absorbed by the skin or clothing, which has a much lower albedo.

Unreflected energy can be absorbed, effectively heating the earth's surface. The surface, in turn, reradiates energy into the atmosphere and space. The Wien displacement law tells us that, at the average temperature of the earth's surface (about 60°F or 15.5°C)—which is, fortunately for us all, much lower than that of the sun—the radiation (*terrestrial radiation*) will be primarily within the longer wavelengths, the infrareds, which are not visible to the human eye. The gases of the atmosphere, as selective absorbers, have the capacity to absorb fairly well in the longer wavelengths, and it is primarily through this indirect process that the atmosphere is heated. This is the so-called *greenhouse effect*. The process is analogous to the free passage of solar energy through the glass of a greenhouse, the subsequent heating of the floor or bottom of the greenhouse, and the trapping of outgoing (longwave) radiation by the glass. Actually, the high temperatures in the greenhouse are also caused by the capacity of glass to prevent heat loss through convective transport.

In the atmosphere, heat can also be acquired by *conduction,* or contact of the lower layers of air with the heated earth surface. *Convection* is a process whereby the heat is transported vertically within the atmosphere, while *advection* describes the horizontal transfer of heat from place to place. Another very important means by which the atmosphere is heated, the release of *latent heat* of condensation, will be discussed more fully in a later chapter.

Summary

There are two ways in which the variation of solar energy within the Great Lakes area constitutes a major control over the weather and climate of that area. First, the location of the Lakes in the middle latitudes dictates that large differences in the amount of solar energy available for surface heating will be seasonally experienced. These differences result from the contrasts between winter and summer in sun angle, length of day, amount of cloudiness, and albedo. As a consequence, the

Great Lakes experience relatively warm summers and cold winters.

Residents of the Great Lakes area take seasonal change for granted. Dress habits change, expenditures for heating fuel or air conditioning fluctuate seasonally, recreational activities reflect seasonal change, and adjustments in living habits and everyday activities follow regular seasonal patterns. Aesthetic appreciations are also often geared to seasonal change. The golden splendor of autumn, the harsh beauty of winter, the soft greenery which is spring, and the mellow warmth of summer—all are images related to seasonal differences in energy availability. How often do we realize that these perceptions of changing seasons are not shared by all people—that over one-third of the world experiences a tropical climate where seasonal change is either absent or perceived in a much different manner?

Secondly, the latitudinal spread of the Lakes, nearly 8 degrees, is sufficient to cause substantial energy reception differences during winter, when the angle of inclination is steeper in the southern parts of the basin and the days are longer. Thus a steep temperature gradient exists over the region during the winter, and the northern portions of the Lakes are significantly colder than the southern portions. During the summer, although the angle of inclination is larger over the southern Lakes, the days are longer in the north. (Remember, the length of day increases with increasing latitude during the summer.) Thus, smaller variations of energy occur between the northern and southern portions of the Lakes, and the temperature gradient is smaller.

2: The Great Lakes and the General Circulation of the Atmosphere

Yearly Variation of Solar Energy over the Earth's Surface: Its Relation to the General Circulation

During an average year, solar radiation amounts received near the equator average about two and one half times greater than those received at the poles, enough to create a heat surplus in the tropics. This is chiefly because the sun shines at a more direct angle within the tropics. At poleward latitudes of about 40 degrees, however, a deficit of energy occurs. The reason why tropical areas do not become warmer and warmer as a result of their yearly energy surplus, and why middle and high latitudes do not become colder and colder as a result of their yearly energy deficit, is that various forms of heat transport occur to offset the imbalance in energy reception. Note that the Great Lakes are latitudinally located at about the boundary between surplus and deficit latitudes and are therefore subjected to rather vigorous energy exchange between the tropics and poles—one reason why the Great Lakes area is an active weather zone.

This exchange of surplus energy is associated with the planetary wind belts, the jet streams, the moving cyclonic systems, and with many other features of the atmosphere which we collectively describe as "weather."

The *general circulation* is the mean motion of the atmosphere which becomes apparent when winds are averaged over long periods of time. The general circulation consists of the surface wind systems that directly affect the Great Lakes in many ways and also air movement at the higher levels, which is important in less obvious ways. While the driving mechanism of the general circulation is the surplus of energy received in the tropics as compared to the middle and high latitudes, the general circulation itself is shaped and molded by the rotation of the earth. The location of the Great Lakes in regard to the weather features associated with the atmosphere's general circulation constitutes a weather and climate control second in importance only to solar energy variation.

The General Circulation and the Great Lakes

If the earth did not rotate and the sun revolved around the earth as the ancients believed, the general circulation would be much simpler to describe. The prevailing winds over the Great Lakes, both at the surface and aloft, would be much different, as would the weather and climate of the region, than they actually are. Near the heated equator, air would expand and rise according to the gas laws, which state that heated air becomes less dense, thus buoyant, and tends to rise.* Aloft near the equator, the winds would flow toward the higher latitudes (figure 6). The air would lose heat on its poleward journey, become denser, and sink near the poles. At the earth's surface, colder air from the poles would flow toward the equator, replacing the warmed air which rose. Circulation within the

*Boyle's law tells us that the volume of a gas is inversely proportional to the pressure, while Charles's law states that the density of a gas is inversely proportional to its temperature. The two laws may be combined into a single equation called the *equation of state*.

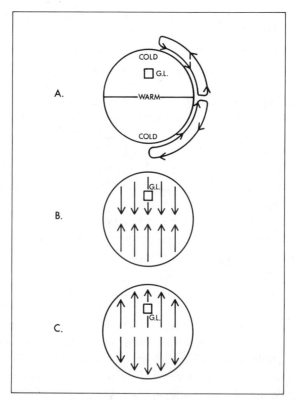

A. Convectional circulation—rising of warm air near the equator, and sinking of cold air near the poles. **B.** Surface winds over the Great Lakes would be northerly. **C.** Winds at upper levels over the Great Lakes would be southerly.

Figure 6   General Circulation of the Atmosphere in Relation to the Great Lakes *(G.L.)*, Assuming the Earth Did Not Rotate

atmosphere would resemble a giant convectional cell, similar to that observed on a smaller scale within a liquid when temperature differences exist. It can be readily determined from figure 6 that the prevailing winds over the Great Lakes would be from the north (winds are always described according to the direction *from* which they blow). Aloft, the prevailing flow of air would be from the south. How simple things would be for the meteorologist!

Effects of Rotation of the Earth

Unfortunately, things are not so simple. The earth, as we know, does rotate, and rotation makes for a much more complicated general circulation. One way to visualize the effects of rotation on the atmosphere is by experimenting with fluids in dishpans. In many ways, the atmosphere behaves like a fluid and responds to the laws of fluid dynamics. If a heating element is placed around the perimeter of the pan (which represents the earth's equator) and a cold source is placed in the center of the pan (representing the pole) one may simulate the atmospheric behavior by rotating the pan. A dye may be introduced so that the motion of the fluid can be observed. Time-lapse photography by a camera rotating with the pan can then produce a record of the fluid motion in the pan.

If the pan is not rotated, the circulation of the fluid resembles the convectional cell we described as occurring in the nonrotating atmosphere. As the speed of rotation is increased, however, a series of waves or eddies appears circling the cold source at the center of the pan. Their number and shape depend on the rotation rate and the heat differential (figure 7). As the waves approach the perimeter of the pan, cold fluid from the center is transported toward the edges of the pan; and as the waves approach the center of the pan, warm fluid is transported toward the center. The deflecting force introduced by the pan rotation will not allow the cell-like convectional circulation to persist. Very simply stated, fluid which flows from the perimeter toward the center of the pan is deflected to the right (with the dishpan rotated in a counterclockwise manner similar to earth rotation when viewed above the North Pole). This deflection is the *coriolis effect,* an apparent force which deflects all freely moving objects (including air parcels) to the right in the

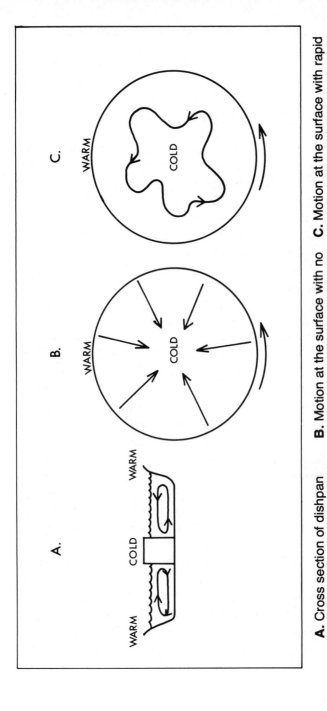

A. Cross section of dishpan showing convectional cells with no rotation.

B. Motion at the surface with no rotation.

C. Motion at the surface with rapid rotation.

Figure 7   Motion of Fluid Associated with Dishpan Experiments

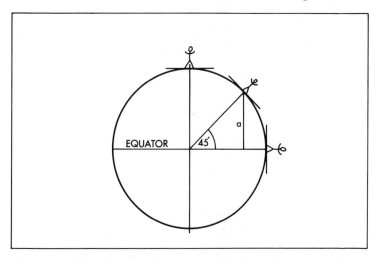

Figure 8   Rotational Components in the Plane Perpendicular to the Earth's Axis (coriolis effect). In the plane tangent to the North Pole, complete rotation occurs every 24 hours. At the equator, no rotation occurs, but "upside-down" motion in relation to the axis exists (coriolis effect absent). At intermediate latitudes (45° N), some rotation and some "upside-down" motion occur; rotation is proportional to the sine of the latitude (a).

Northern Hemisphere and to the left in the Southern Hemisphere.

How closely do the dishpan experiments simulate the behavior of the earth's atmosphere? One shortcoming is obvious. The atmosphere is a spherical shell which surrounds the earth. The dishpan, however, is flat. Thus when the dishpan is rotated, all parts of the pan turn through the same angular distance in respect to an axis perpendicular to the center of the pan (figure 8). However, in the real atmosphere, only those points at the same latitude turn through the same angular distance in a plane perpendicular to the earth's axis. Points near the tropics undergo little rotation in respect to the earth's axis, while points near the pole undergo maximum rotation. Essentially, then, the nonrotated dishpan gives us a picture of circulation in the tropics, where the rotational component in a plane perpendicular to the earth's axis is nonexistent or very small. With rapid rotation, the fluid in the dishpan behaves like the atmosphere in the middle and high latitudes, where the coriolis effect is much stronger.

Winds of the Tropics

In the earth's atmosphere (figure 9), the tropical cell-like circulation extends poleward from the tropics to the subtropics. Beyond the subtropics, the coriolis effect is so strong that the circulation cannot maintain itself, just as the convectional circulation in the dishpan disintegrated when the pan was rotated more rapidly. Thus the rather simplified circulational pattern which might occur on a nonrotating globe is actually confined to the tropical latitudes. Within the tropical cells, which are called *Hadley cells,* air flows toward the heated equator as *trade winds* from the northeast in the Northern

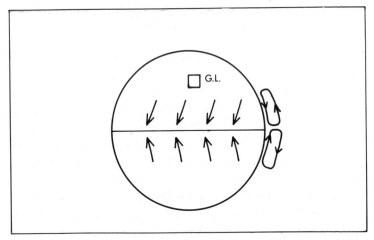

Figure 9   Tropical Circulation Cell and Surface Winds (Trades), Assuming a Rotating Earth

Streamers of cirrus mark a high-altitude jet stream. Cirrus clouds are composed of ice crystals and commonly occur near jet streams.

Hemisphere and from the southeast in the Southern Hemisphere. At the equator, the air rises, a return flow aloft carrying it poleward. In the subtropical latitudes, a pileup of air coming from the equator occurs; as a result of this convergence, the air undergoes a sinking motion or subsidence. Because of the pileup of air aloft at the subtropical latitudes, the weight of the air increases in that zone, and the pressure beneath the subsiding air is high. An outflow near the surface compensates for the accumulation by directing air toward the equator and also poleward into the middle latitudes. The tropical circulation cell can be described as a *meridional circulation* in the mean, as it is aligned primarily parallel to a meridian of longitude.

Winds in the Middle Latitudes:
The Jet Stream over and near the Great Lakes

Over the Great Lakes, however, a much different circulation pattern occurs, one characterized by the transport of heat poleward in eddies and by a wave-like pattern with a jet stream at its core. This is the middle-latitude component of the general circulation which is modeled by the dishpan at rapid rotation. A high-velocity jet stream circles the globe from west to east at the higher altitudes (about 35,000 feet) and is frequently present aloft in the vicinity of the Great Lakes. At the surface, winds are more variable, but flow generally from a westerly direction.

The jet stream flows from west to east, often in a wave-like pattern similar to the behavior of the fluid in the dishpan, transporting warm air northward in its poleward intrusions and cold air southward as it swings toward the equator. Wind velocities near the core of the jet may, in some locations, exceed 150 knots. Above, below, and on the flanks of the jet, wind velocities decrease markedly, and *velocity shears* are said to exist.

The configuration and intensity of the jet stream fluctuate both seasonally and over shorter time periods. Sometimes the jet flows from west to east in a straight-line fashion. At other times, the jet meanders far to the north and south in waves. It is during this latter stage that the jet becomes a prominent mechanism for surplus heat transport from the tropics.

In winter, when the temperature contrast between the poles and the equator is very strong, the jet stream is a powerful feature. It attains its highest velocities at this time of the year and forms an expanded vortex extending toward the equator. In the summer, when the temperature contrast is much smaller, the jet is much weaker, and is positioned closer to the pole.

3: How the Jet Stream Affects Surface Weather

HOW DOES THE JET STREAM AFFECT THE WEATHER and climate of the Great Lakes? After all, the jet is a powerful current of air at levels 30,000–35,000 feet above the surface. Unless we are pilots or passengers in high-flying aircraft, we will probably never directly encounter a jet stream. But its indirect effects are very strongly felt in various manifestations of surface weather, particularly in the Great Lakes region.

Fronts and Air Masses

Beneath the jet near the earth's surface, warm air masses from the south and cold air masses from the north collide along a *front* or boundary zone between air currents with unlike temperatures.* Nowadays, the symbols used by weathercasters to represent fronts near the earth's surface are familiar to many and we speak of cold fronts and warm fronts with a glibness which belies the fact that the concept is a relatively recent contribution. Scandinavian meteorologists established a dense network of weather stations during World War I. The information from this network allowed the Scandinavians to develop

---

*Actually a number of jet streams occur within the atmosphere. The jet referred to here is the *polar front jet,* and it is this one which is instrumental in governing weather over the Great Lakes.

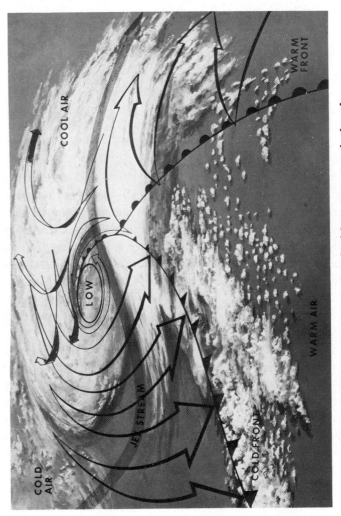

Fronts and air masses associated with an extratropical cyclone.

some new concepts about weather, as detailed information from the lower atmosphere became available for the first time. They recognized the existence of a front as an inclined surface separating two atmospheric portions unlike in temperature, humidity, or both.

The concept of the *air mass* was another contribution of the Scandinavians. Air masses were defined as large portions of the atmosphere which were rather uniform in temperature and humidity. They form, it was found, when the atmosphere stagnates for long periods of time in certain areas and thus takes on the characteristics of the earth's surface over which they reside. For example, if a portion of the atmosphere languishes for a long time over the vast snow-covered Arctic tundras in the winter, that portion could become very cold and dry, like the surface. If a part of the atmosphere stagnates over the warm tropical oceans, it becomes very warm and moist. Air masses can then be transported from their original birthplaces (or source regions) to other areas; and although constantly being modified and changed in their travels, they are capable of conveying the original characteristics of their source regions to distant localities.

The air mass types which typically affect the Great Lakes reflect the region's situation midway within the open corridor that extends from the Arctic to the subtropics. Free interchange of air masses from tropical and polar origins is thus permitted. On the other hand, the presence of high mountain ranges along the western margin of North America prevents an easy incursion of air masses from that direction; and, with the prevailing flow from the west of air aloft, air masses from the Atlantic are carried downstream toward western Europe.

Air which streams southward over the eastern part of the United States and over the Great Lakes area is usually cool and dry (figure 10). This air is sometimes called a Canadian air mass because its source region is the vast northern expanse of Canada. The meteorologist calls this type of air mass *continen-*

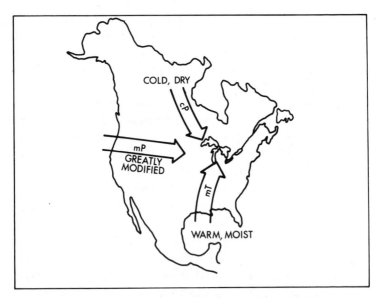

Figure 10   Air Masses Affecting the Great Lakes

tal Polar (*cP*), indicating its formation over a land or continental surface in the polar latitudes. In winter, this air mass can be extremely cold, originating in far northern Canada over snow-covered terrain, in which case the weather analyst often refers to it as an *Arctic air mass*. In summer, when the Canadian north is free of snow, this air mass is, of course, much less cold than in winter and may be described as just cool. The Great Lakes lie athwart the main travel path of these air masses as they move from their source region toward the warmer latitudes.

At the other end of the corridor is the Gulf of Mexico and the southern portion of the North Atlantic Ocean. These are warm regions within the global zone of energy surplus; being water surfaces, they are capable of supplying large amounts of moisture to the overlying atmosphere. The meteorologist designates air masses forming here as *maritime tropical* (*mT*) according to the source region, a tropical ocean. In the summer, this air mass brings heat and humidity into the higher latitudes, and when the air mass reaches as far north as the Great Lakes, the area may experience a period of uncomfortable weather. In the winter, an intrusion of mT air into the Great Lakes may cause a thaw or a brief interlude of abnormally high temperatures.

It should be stressed that this constant exchange of unlike air masses over the Great Lakes, made possible by meanderings of the jet stream and the geography of North America (which provides a lowland corridor in the eastern portions), is a major reason for the very active weather which occurs within the Great Lakes area.

Cold Spells and Warm Spells

The jet stream's wave-like meanderings poleward and equatorward determine the occurrence of weather *spells,* warm or cold. These are short-term, nonperiodic weather patterns superimposed on the progression of the seasons. We know, for example, that as spring approaches and more solar energy is received, there is a general upward trend in temperature. But the upswing of the thermometer does not uniformly occur. Cold spells follow warm spells, and there are occasional regressions into periods when winter returns in all of its harshness. But if the irregularities are evened out, a warming trend becomes apparent. The irregularities are responses to changing positions of the jet stream. When the jet swings north of the Great Lakes, a warm spell ensues as tropical air masses move northward, and when the jet tracks south of the region, it draws cold air masses southward from Canada (figure 11). It is interesting to note that the jet's position may persist for a considerable period of time—weeks, months, occasionally even for an entire season. Such was the case during the very cold winter of 1976–77, when the jet stream maintained an aberrant position for more than 4 months. It is the mean position of the jet over a given time

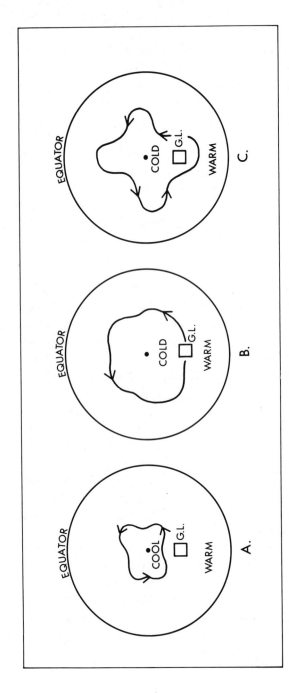

A. Typical summer situation with the jet located north of the Lakes.

B. Typical winter situation with prevailing west-to-east flow (high zonal index).

C. Typical winter situation with poleward loops and equatorward meanders (ridges and troughs; low zonal index).

Figure 11 Varying Positions of the Jet Stream in relation to the Great Lakes

period which is largely responsible for the occurrence of either abnormally cold or abnormally warm weather.

We should also consider another possible jet alignment. Suppose the jet stream flows over eastern North America directly from west to east without any pronounced loops and meanderings. Under these circumstances (which the meteorologist describes as *westerly* or *zonal flow*), neither the Canadian air masses nor the air masses from the Gulf of Mexico are likely to affect the Great Lakes. Instead, air masses of Pacific origin become dominant. These are designated as *maritime polar* (*mP*), indicating a source region over the high latitudes of the oceans, in this case the North Pacific. Obviously, this air mass must move a very long way from its source region before it can reach the Great Lakes. During its trip from the Pacific Ocean, the air mass is highly modified. It must ascend and descend mountain ranges, losing moisture as it is forced to rise on the windward slopes and warming as it descends the leeward slopes. It may reside for awhile over the Great Basin between the Rockies and the Sierra-Cascade ranges, then travel a great distance over the prairies toward the Midwest. Modifications occur all along its travel route. As a result, when Pacific air under westerly flow aloft reaches the Great Lakes, its Pacific origin is barely recognizable. It may bring mild weather to the region without the temperature extremes associated with the polar and tropical air.

In summary, when the jet is looping northward over the eastern United States and is positioned north of the area, warm spells influenced by warm and humid air masses from the Gulf of Mexico are likely to occur. When the jet is looping southward over the eastern United States or is positioned far to the south, cold spells may ensue. A straight westerly flowing jet is likely to bring modified Pacific air to the Great Lakes along with a tendency toward "normal" weather and no temperature extremes. With these considerations in mind, we can see that the jet stream is an important feature in determining the spells of weather which are so common in the Great Lakes.

More About Fronts:
Their Nature and Relation to the Jet Stream

As mentioned previously, a front extends downward from the globe-circling jet to intersect the surface of the earth, separating warm tropical air from cold polar air. Thus the front is a zone of sharp temperature contrast, separating unlike air masses.

If the earth did not rotate, the behavior of the atmosphere at a frontal zone would resemble that of two fluids having different densities in a container. The heavy fluid would sink to the bottom, and the lighter fluid would rise to the top (figure 12). In the atmosphere, the warmer, less dense air would rise to the top, and the heavier, colder air would sink to the bottom. The boundary line between the unlike air masses would be parallel to the earth's surface.

The rotation of the earth complicates atmospheric processes by introducing the coriolis effect. Rotation creates a boundary surface which is inclined to the horizontal and intersects the earth's surface at an angle. Warm air rises along the inclined plane, while cold air wedges beneath. The intersection of the inclined boundary surface with the earth's surface is represented on weather maps by a special type of symbol, according to whether the cold air mass or the warm air mass is advancing (figure 12).

If the cold air mass is winning the battle and advancing into areas formerly occupied by warm air, a *cold front* is said to exist. Frictional drag exerted from the roughness of the earth's surface retards the movement of the cold air in its lower layers. Thus the inclination of the boundary surface between the air masses becomes quite steep, with a slope as much as 1/100 (meaning that for every 100 horizontal miles—about the distance from Chicago to Milwaukee—the surface slopes

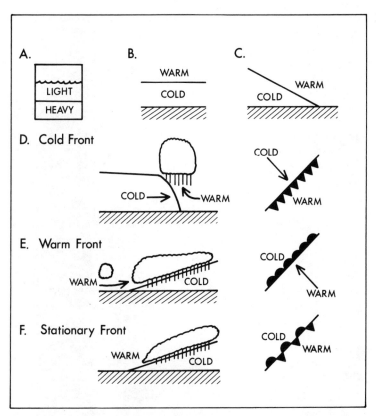

A. Container with fluids of different densities. **B.** Nature of frontal boundary surface with nonrotating atmosphere. **C.** Nature of frontal boundary surface with rotating atmosphere. **D.** Cold front in vertical representation and representation on a surface map. **E.** Warm front in vertical representation and representation on a surface map. **F.** Stationary front in vertical representation and representation on a surface map.

Figure 12   The Nature of Fronts

upward 1 vertical mile.) Where warm air is replacing cold air, a *warm front* exists. The slope of the boundary is more gradual, perhaps 1/300—about the distance in miles from Detroit to Sault Ste. Marie, Michigan.

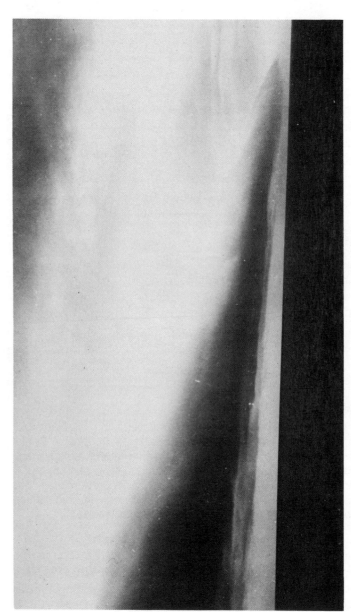

A thunderstorm gust front. Severe weather may occur in advance of or near cold fronts.

Because of the difference in slopes along the inclined boundaries, the weather sequence accompanying the presence and passage of cold and warm fronts is also somewhat different. With the cold front, rapid temperature change from warm to cold, wind shifts, and occasionally violent weather features are likely to occur, but weather events are usually not long-lived. The violent weather, including the possibility of severe thunderstorms and tornadoes, may occur because the warm air being replaced is pushed upward sharply. We will see later that *uplift* is necessary to cause cloud formation and precipitation—and the more rapid the uplift, the greater the likelihood of severe weather occurrences. Along the warm front, however, the warm air slides gradually upward along the boundary plane. Although the weather sequence associated with a warm front may be long-lived and cover a large area, the likelihood of severe weather is not so great as with a cold front.

Because of the jet stream positioning over or near the Great Lakes area throughout a large portion of the year, the region is subject to many fronts. This means that rather sharp temperature differences within the region are not uncommon. Occasionally, in fact, temperatures may vary as much as 40 or even 50 degrees. These differences, too large to be caused by radiation differences, result from the presence of two contrasting air masses separated by a front. In addition to the large temperature contrasts which may occur, the Great Lakes area also experiences the active weather associated with frontal zones.

Cyclones and Anticyclones

The jet stream affects the weather of the Great Lakes in yet another very important way. In describing the general circulation of the earth's atmosphere, we stated that air near the heated equator tends to rise, while air at the subtropics descends within the poleward flowing branch of the tropical circulation cell. We have not yet mentioned the differences in atmospheric *pressure* which exist over the surface of the earth.

Pressure can be defined as the weight per unit area of an air column extending to the top of the atmosphere. This pressure or weight is the combined pressure of the gases which make up the atmosphere—nitrogen, oxygen, argon, carbon dioxide, ozone, water vapor and the host of minor gases. Pressure would decrease with height quite rapidly within the denser air layers nearest the earth's surface and less rapidly in the higher layers. By the time one ascends to the 18,000-foot level, about half of the atmospheric pressure would be below.

Pressure can be measured by *barometers* and expressed in units of inches of mercury, or in millibars.* The millibar (mb) is preferred by the National Weather Service, but inches of mercury is the customary expression used on television, radio, and in newspapers. At the earth's surface, pressure varies with time and from place to place. The actual amount of this variation is not large, but it is extremely important. Areas of relatively low pressure are called *lows* or *cyclones,* while regions of high pressure are called *highs* or *anticyclones*. Pressure differences at the earth's surface cause the winds and control their direction and velocity; for this reason, pressure is an important atmospheric control.

The gas laws tell us that air which is heated becomes less dense. As mentioned in our general circulation discussion, the heated air near the equator tends to rise because of lessened density or pressure. We generally find that an area of low pressure, accompanied by rising air, exists near the equator. On the other hand, near the poleward limits of the tropical circulation cell (subtropics) where air begins to pile up, pressure at the surface becomes higher because of this accumulation of air aloft; and the air begins to sink or subside within the air

*The American Metric Council has proposed that after 1 January 1979, pressure will be expressed in *kPa* (kilopascals). A kilopascal is equal to 0.10 millibars. The National Weather Service, however, has postponed its planned metric conversion timetable.

column. At the bottom of the air column, air flows outward toward regions of lower pressure.

From the example of the tropical circulation cell, we can form some general principles. At the surface, where pressures are high, air generally flows outward, and a downward motion occurs. Where pressures are low, air flows inward and rises. All of this movement results from the tendency of the atmosphere to equalize the pressure imbalances which exist from place to place. These pressure imbalances can exist because of temperature (density) differences, as in the case of the low pressure area near the equator; or because of the accumulation (or depletion) of air at the higher altitudes, as in the case of the subtropical highs. In the Great Lakes region, we are primarily concerned with a large number of *mobile pressure systems* which pass through the area, each bringing sequences of weather accompanied by changes in the wind field.

How the Jet Stream Can Cause Surface Cyclones and Anticyclones

We must next examine the ways in which the jet stream, flowing thousands of feet above the surface, can influence the formation of mobile high and low pressure centers. As the jet stream meanders about the Northern Hemisphere, it accelerates as it approaches regions known as *jet cores* and decelerates as it exits these areas. Also, the jet sometimes swerves toward the right (as when it completes a poleward loop and begins an equatorward loop), and sometimes swerves to the left (as when it is at the apex of an equatorward loop and begins to swing poleward). Consider also that wind velocities laterally decrease rapidly to the north and south of the jet stream, causing strong velocity *shears*.

All of these things—accelerations and decelerations, the curved path of the jet, and velocity shears on the flanks of the jet—can either cause air to accumulate (converge) or spread apart (diverge) in the area of the jet. Although these effects may

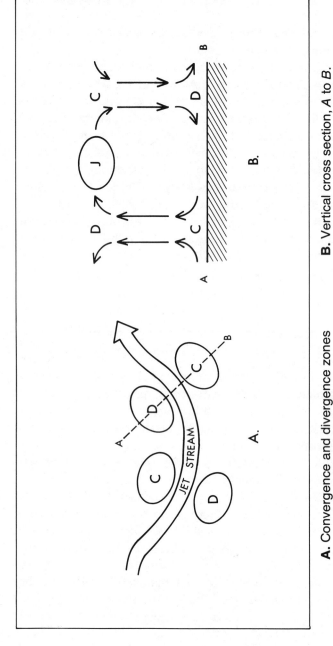

A. Convergence and divergence zones on the flanks of the jet stream.

B. Vertical cross section, *A* to *B*.

Figure 13 Initiation by the Jet Stream of Upward and Downward Movements in the Atmosphere

occur at higher atmospheric levels, they are accompanied by compensatory effects at the surface.

Figure 13 shows zones of convergence and divergence established along the flanks of a high altitude jet as it makes an equatorward loop in the Northern Hemisphere. A cross-sectional view through points *A-B* indicates that beneath the divergence zone in the left front portion of the jet, air converges and rises. In the right front portion of the jet, air converges. Compensating at the surface, downward moving air spreads out or diverges. Beneath the zone of upper air divergence where a column of air is being depleted, pressures fall. Pressures rise beneath the zone of upper air convergence. In this way, the jet, as it meanders along at high levels, can establish surface zones of low pressure, or cyclones, and high pressure, or anti-cyclones.

Summary

The ways in which the locations of the Great Lakes and the jet stream, as a component of general circulation in the middle latitudes, interact include the following: (1) located halfway between the North Pole and the equator, the Great Lakes frequently experience the jet stream, either directly aloft or nearby; (2) in summer, the jet stream is normally weak and located north of the Great Lakes; (3) in winter, the jet intensifies and is normally positioned south of the region; (4) spells of abnormally warm or cold weather in the Great Lakes are related to poleward or equatorward meanderings of the jet; (5) fronts are frequently found at the earth's surface within the region as a result of their association with the jet stream aloft; and (6) mobile cyclones and anticyclones are common features in the Great Lakes area, resulting from divergence and convergence in the upper air caused by the jet, with compensating horizontal and vertical motions near the surface.

4: Cyclones in the Great Lakes Region

The Wave Theory of Cyclone Formation

Another way by which cyclones can form proceeds from the bottom up—that is, events which first occur at the surface may eventually cause a low-pressure area to develop. At this point, one may wonder just how cyclones form—from the upper air downward or from the surface upward? The answer: in both ways.

Before anyone had ever heard of jet streams, the Scandinavians coped with the problem of *cyclogenesis,* or the birth of cyclones. Lacking much upper-air data, their major emphasis focused on processes occurring near the surface. Along with the idea of the air mass and the front came the concept of the *wave theory* of cyclone development. The wave theory requires a preexisting front, the formation of a wave along it, and the subsequent development of a low-pressure area or cyclone. The cyclone may then go through a sequential life history consisting of stages—from youth through maturity and ultimately into a dissipating or old-age stage. The life span of a cyclone may range from a day or two to more than a week.

Figure 14 shows a wave cyclone at typical stages of development, following the classic Norwegian model de-

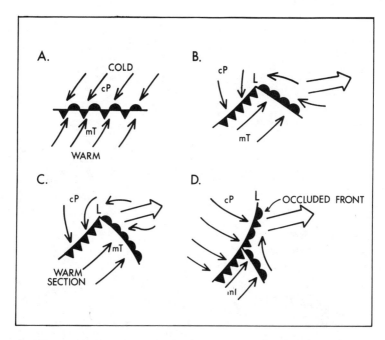

A. Warm winds meet cold winds along a stationary front. **B.** A wave develops on the front. Warm air advances northeastward, cold air southeastward. This is the youthful stage of a counterclockwise circulation. For a cyclone to maintain itself, proper flow aloft must provide a divergence mechanism for converging surface winds. Pressure falls at the inflection point on the wave. **C.** Mature or open state—Jet stream aloft has provided the necessary divergence. The cold front is advancing more rapidly than the warm front, while the warm sector has higher temperatures as mT air at the surface is confined to the south quadrant of storm. Pressure continues to fall at the low-pressure center. **D.** Occlusion stage—Cold front has overtaken the warm front. The cyclone may begin to weaken, or a new cyclone may form at a meeting of cold and warm fronts.

Figure 14   Stages in the Life of a Wave Cyclone

veloped during and after World War I. While the cyclone progresses through its life stages, it may move a thousand or

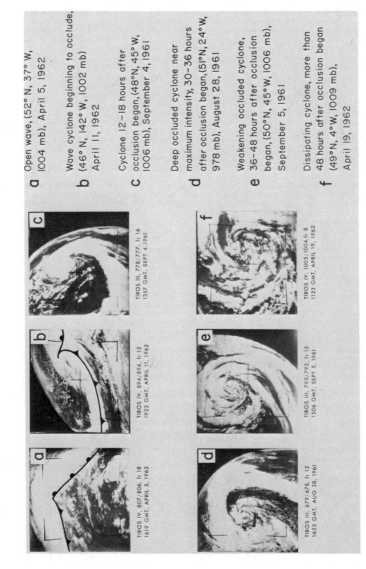

a Open wave, (52° N, 37° W, 1004 mb), April 5, 1962

b Wave cyclone beginning to occlude, (46° N, 142° W, 1002 mb) April 11, 1962

c Cyclone 12–18 hours after occlusion began, (48°N, 45°W, 1006 mb), September 4, 1961

d Deep occluded cyclone near maximum intensity, 30–36 hours after occlusion began, (51°N, 24°W, 978 mb), August 28, 1961

e Weakening occluded cyclone, 36–48 hours after occlusion began, (50°N, 45°W, 1006 mb), September 5, 1961

f Dissipating cyclone, more than 48 hours after occlusion began (49°N, 4°W, 1009 mb), April 19, 1962

Stages in the life cycle of a Northern Hemisphere cyclone.

more miles, guided by the flow of air aloft. Although forming at the surface, it must necessarily incorporate a jet stream into its upper-air structure, as the jet must remove the air converging at the surface by providing for the proper divergence zone aloft.

Many cyclones passing through the Great Lakes area form as a result of conditions aloft (a favorable jet alignment) which have caused surface convergence. Because cyclones are convergent weather systems, they then incorporate fronts into their structure. The open corridor in eastern North America provides an easy means for the further convergence of warm and cold air masses near the center of the storm. Consequently, storms in the eastern United States can become very vigorous and can be accompanied by sharp temperature changes.

But many cyclones also seem to form along preexisting fronts over the United States; they bring in jet streams to sustain them, and they complete their life histories after moving through the Great Lakes area. Which method of formation is legitimate? Both. Remember that because of the location of the Great Lakes, the jet stream is frequently overhead; a large number of fronts occur within the region and upwind; and an easy convergence of warm and cold air masses is possible. All of these factors point to a high frequency of vigorous cyclonic activity over the Lakes—one reason for the lamentable weather about which the early settlers complained, and one reason also for a long record of shipwrecks and Great Lakes disasters. We will now examine cyclones more closely to see why they may bring bad weather as they approach.

Winds in Cyclones and Anticyclones

The existence of pressure differences causes and controls the winds. Wind direction and velocity is determined primarily by the alignment of the *pressure gradient,* or the rate and direction of pressure change. Air, as it moves to equalize the pressure imbalance within the atmosphere, flows from high pressure to low pressure at velocities proportional to the pressure gradient. Again, things would be much simpler for the meteorologist if the earth did not rotate. Air would flow directly from high pressure to low pressure and that would be that. But the old bugaboo of the coriolis effect enters the picture, deflecting air currents to the right in the Northern Hemisphere and to the left in the Southern Hemisphere. The parallel lines in figure 15 are *isobars* drawn through points of equal pressure. At the upper levels (above about 2,000 feet), the pressure gradient force and the coriolis effect are nearly in balance, so

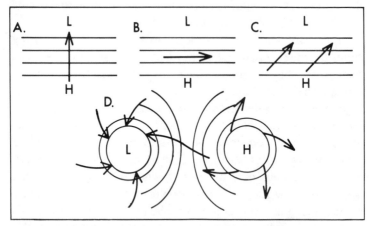

A. Were it not for the earth's rotation, air would flow directly from high pressure to low pressure. **B.** With rotation and the coriolis effect, a balance between the pressure gradient force and coriolis effect is attained in the upper air. Winds flow parallel to the isobars. **C.** At the surface, friction lessens the coriolis effect; winds flow from high pressure to low pressure at an angle to the isobars. **D.** Resulting wind patterns around high-pressure and low-pressure areas.

Figure 15   Winds in relation to Pressure Patterns in the Northern Hemisphere

that the winds flow parallel to the isobars. At the surface, however, friction induced in the lower layers of the atmosphere by rough terrain, obstacles, buildings, and other surface irregularities reduces the wind velocity. As the coriolis effect is partly a function of velocity, the reduction of velocity due to friction lessens the coriolis force. The net result is that a balance is not obtained. The air does not flow parallel to the isobars but at an angle across the isobars from high pressure to low pressure.

So far, our illustration has considered isobars as straight lines which are parallel to each other. In the atmosphere, however, the pressure systems are more often defined by circular or concentric isobars. Thus the flow of air around high-pressure and low-pressure areas at the surface in the Northern Hemisphere can be described as follows: Air flows out of high-pressure areas in a clockwise fashion and into low-pressure areas in a counterclockwise fashion (figure 15).

As high-pressure and low-pressure areas guided by the flow of air at the upper levels pass through the Great Lakes area, they behave much like small whirlpools or eddies within a river. The river of air, with the jet stream at its core, flows generally from west to east. But as the whirlpools and eddies at the bottom of the air river pass by an observer, their individual wind circulations may cause winds from varying directions. Thus, although the higher-level winds of the Great Lakes area flow predominately from the west, the surface winds may be much more variable, responding to the tracks of individual cyclones and anticyclones as these are formed and steered by the jet. Since sharp pressure gradients often accompany the passage of cyclones, the wind velocities may occasionally reach destructive levels.

As the winds converge horizontally near the center of cyclones, vertical compensation occurs. The rising motion or uplift which accompanies cyclones is the chief reason why clouds and precipitation commonly accompany their passage. We will explore the reasons why this is true in a later chapter.

Frequencies of Cyclones and Anticyclones

The frequency with which traveling cyclones and anticyclones visit the Great Lakes area varies seasonally, as does the mean position of the jet stream over the eastern United States. As the temperature gradient between the North Pole and the equator lessens in the summer, the jet stream weakens and retreats northward into central Canada. August normally brings a minimal amount of cyclonic activity over the Great Lakes, although cyclones may be very frequent across central Canada. With the jet over central Canada during summer, high-pressure areas are more numerous over the Lakes. In addition, the physical properties of the Lakes at that season favor the intensification of anticyclones, a fact discussed in a later chapter.

In the fall, the temperature contrast between the polar regions and the tropics sharpens, and the jet stream begins to intensify and swing southward. Cyclones become more frequent, particularly in the northern portions of the Lakes. Anticyclones are less frequent, partly due to the role of the Lakes themselves, which tend to attract cyclones and weaken anticyclones during the fall and winter months. Maximal cyclone frequencies occur in December and January, since the jet stream at that season is at its most intense equatorward displacement. Winter anticyclones mostly form over the northlands of Canada. When they slip southward, they normally enter the United States in the vicinity of the Dakotas, plunge southward, then veer sharply to the east, traveling across the middle South toward the Atlantic. Anticyclones may also slip across Canada north of the Great Lakes, move across New England, then out to sea. Few anticyclones move directly over the Great Lakes; and it is a rare occasion when a weather map shows an anticyclone positioned directly over the Lakes during winter.

With the approach of spring, the jet stream begins its retreat northward over eastern North America from its most southerly mean position (south of the Great Lakes) to its summer track across central Canada. Low-pressure areas then tend to move farther and farther north. They may be fairly frequent in the northern Lakes area during late spring or early summer, but are weak and infrequent over the southern Lakes. Anticyclones occur more often, and the weather map may show them positioned directly over the Great Lakes.

Seasonal progression of the jet southward over the Lakes as winter approaches and the increased number of low-pressure areas passing through the Lakes area is one reason why. Great Lakes winters are cloudy and gloomy with many storms, while the summers are much sunnier. There are other reasons too, stemming from the interactions of the Lakes themselves with the atmosphere.

Tracks of Cyclones and Their Behavior in the Great Lakes

Cyclones exhibit preferences for certain patterns of travel, and a quick glance at figure 16 tells us that the tracks tend to converge over the Great Lakes and New England. Consequently, these regions tend to be very stormy with quick changes of weather.

Low-pressure areas which pass through the Lakes tend to be accompanied by some rather typical weather sequences which partly, at least, depend on where the cyclone was born. Probably the most common of all cyclone types is one which originates in Alberta, Canada, just to the lee of the Rockies. This type of storm occurs most frequently in the fall or early winter. With the jet stream looping southward over the eastern United States, Alberta cyclones tend to move southeastward

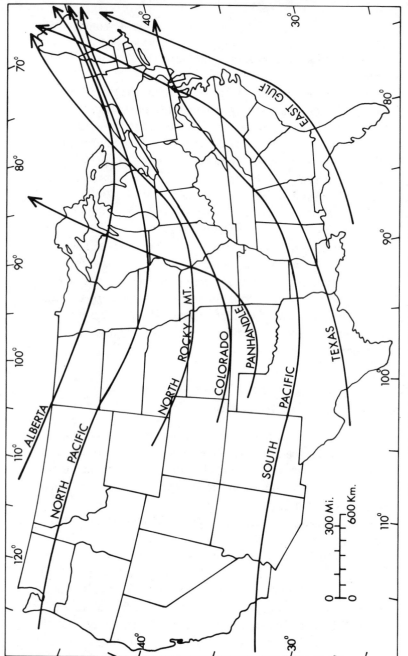

Figure 16   Main Tracks of Cyclones in the United States

out of Canada, recurving toward the east of northeast over the Lakes, then speeding eastward up the St. Lawrence valley or over eastern Canada. They are sometimes referred to as *Alberta clippers* because they move at 40 or 50 miles per hour. They also tend to occur in families, so that as soon as one Alberta storm has crossed the Lakes, only a brief respite intervenes before the next one approaches. If the flow of air aloft remains constant for a long period of time, storm after storm may appear to the northwest of the Lakes, drop southeastward, and recurve over the Lakes in a repeatable (and sometimes enervating) weather sequence. Such was the case from November to February of 1976–77, when winter held the Lakes for weeks on end without letup or thaw.

In most cases, Alberta lows recurve quickly so that they cross only the northern Lakes. Occasionally, however, the jet stream may carry them farther south, causing them to recurve in lower latitudes and travel to the south of the lower Lakes.

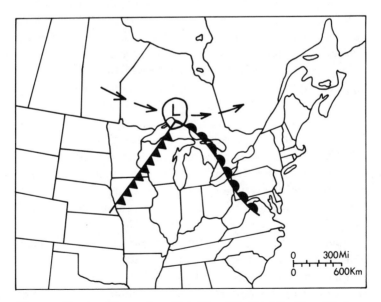

Figure 17   An Alberta Low Tracking North of the Great Lakes

The weather effects associated with the Alberta type of storm are amplified in the higher latitudes, which are, in most cases, closer to the storm track. Given the counterclockwise inward circulation around the approaching cyclone, surface winds generally turn southeast as the low-pressure area approaches, veering to the south and southwest as the storm recurves through, or north of, the Lakes (figure 17). (To *veer* is to shift in a clockwise direction; to *back* is to shift in a counterclockwise manner.) In a situation where the storm track is south of the Lakes, winds may back from easterly to northeasterly and then into the north.

Although Alberta cyclones occur with high frequency, their tracks in the more northerly latitudes do not bring them in proximity to a major moisture source. The major moisture source for the Great Lakes is the Gulf of Mexico. Consequently, Alberta cyclones generally do not produce large amounts of precipitation, although during winter they may be capable of causing moderate snows over the upper Lakes, particularly in lee shore and elevated regions. As these storms generally recurve through the upper Lakes or in southern Canada, the more southerly portions of the Lakes experience cloudy skies, moderating temperatures with light snow or rain, and a change to colder weather as the accompanying cold front passes. The Alberta cyclone usually does not bring severe storm conditions, so meteorologists watch other areas for forebodings of vigorous and dangerous storms.

When the jet stream loops southward over the western United States and northward over the East, cyclones may form over the high plains area of Colorado, Oklahoma, or Texas. These cyclones often become quite intense; as they travel nearer to the Gulf of Mexico, they may be capable of producing large amounts of precipitation. After intensification, the jet stream aloft steers them northeastward toward the Great Lakes, and that area braces for a bout of bad weather. The exact weather sequence is largely determined by the subsequent path of the storm. If the storm travels northeastward so that its center passes *north* of the observer's location, the weather sequence is characterized by warming southerly winds bringing maritime tropical air from the Gulf of Mexico (figure 18). Midwinter thaws or mild periods are likely to accompany this circulation pattern, particularly in the southern sections of the Great Lakes. During spring or autumn, warm spells with rain and thundershowers may be typical. Winds initially blow from the east or southeast as the cyclone approaches, veering into the south and eventually into the southwest and west as the storm center passes to the north. The fronts accompanying the cyclone, characterized by windshifts and sharp temperature changes, will also pass through the Great Lakes. South and east of the storm's path, warm air from the south is brought northward by the convergent circulation around the center of the low. To the north of the storm path, temperatures remain cold, and winds back from northeast to northwest. Snow may also occur along a zone north and west of the storm's path.

Should a storm of this type be steered along a more southern track and pass *south* of the Lakes, the weather sequence is quite different. Warm air at the surface remains south of the storm track, and frontal passage does not occur. At the height of winter, when temperatures are quite cold, precipitation is primarily in the form of snow, with heaviest amounts in a band oriented southwest-northeast about 100 to 150 miles north of the storm track. Freezing rain or sleet may occur in a narrow band just to the south of the snow zone. South of the zone of freezing rain will be rain. In the warm sector of the storm, thundershowers may break out in advance of the cold front (figure 19).

Although low-pressure areas may occasionally approach the Great Lakes from almost due south and may even plummet directly southward from Canada, these are unusual situations. Much more commonly they follow one of the two basic tracks outlined. It should be evident from the foregoing discussion that

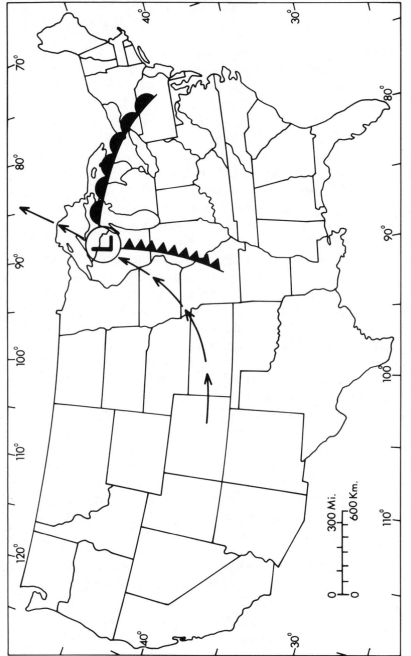

Figure 18 A Colorado Low Tracking North of the Great Lakes

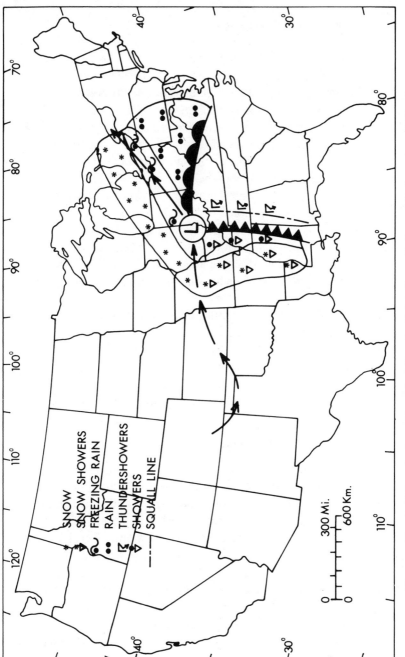

SNOW
SNOW SHOWERS
FREEZING RAIN
RAIN
THUNDERSHOWERS
SHOWERS
SQUALL LINE

300 Mi.
600 Km.

Figure 19 A Colorado Low Tracking South of the Great Lakes with Associated Weather

the exact weather sequence for any given locality in the Great Lakes area will much depend upon whether the center of the disturbance passes north or south of the observer. In the summer, of course, most cyclones are traveling along the northern path across central Canada, so that an observer in the Great Lakes area will be mostly south of the path. In the winter, however, cyclones may pass either north or south of the Lakes. As an example, for weeks in the autumn of 1975 low-pressure areas formed over the central prairies and swung northeastward into Canada, passing north of the Great Lakes. The major control of this track was the positioning of the jet stream aloft. Because of the jet's position, warm air was continuously brought northward into the Great Lakes region and polar air was kept far to the north in Canada. An unusually warm fall was the result.

Around mid-December, the jet and storm track shifted to the south. Alberta clippers came flying out of the northwest, recurving over and sometimes to the south of the Lakes. Cold air was brought southward by this change of circulation. Since many of the storms went south of the area, very little warm air from the Gulf could interrupt the cycle of polar air control.

Some Forecasting Tips for Great Lakes Residents

The approach of a cyclone is usually a well heralded event in this era of modern communications and scientific forecasting techniques. In most cases, forecasters know well in advance its general direction of approach to the Great Lakes. Adequate foreknowledge of the exact track of the storm is critical for accurate forecasting. But this is one of the most difficult forecasting problems. Fortunately, some simple rules of thumb exist which will help the layman predict the approximate track of the approaching storm.

Many people own or have access to barometers. A rapidly falling barometer certainly gives excellent indication that an

Wind direction, when watched carefully, may provide a clue to future weather.

area of low pressure is approaching; in the Lakes area, it is a sure bet that a cyclone is threatening from the southwest, west, or northwest. But from which of these directions? The wind gives an excellent clue. Any shift of the wind to an easterly quadrant, in fact, is in itself a harbinger of the approach of bad weather. Sometimes east winds do bring fair weather to the Great Lakes, but not very often.

The wind direction should then be watched very carefully; also watch the appearance of the sky. If the winds begin to back to the northeast, it is sure that the storm is approaching from the southwest and will pass south of the observer. Along with this track, a gradual thickening of clouds will occur; temperatures will remain steadily cool or cold; precipitation of extended duration will occur, possibly becoming heavy; and the barometer will continue to fall until the storm center is at its closest point to the observer. Then the barometer will gradually rise, winds will back to the northwest, and temperatures will slowly fall as the cyclone passes to the east.

If, however, the wind vane shows a tendency for the winds to veer (shift in a clockwise manner) to the southeast, the cyclone is approaching from the west or northwest and will pass north of the observer. Skies will become increasingly overcast, and rain or possibly snow will begin. As the warm front which accompanies the cyclone reaches the observer, winds will veer suddenly to the south or southwest; skies may clear or become partly cloudy; and the temperature will rise under the influence of maritime tropical air from the Gulf of Mexico. The warm period is short-lived, however, and as the storm continues to move eastward, eventually the cold front will reach the observer. The rapid, forced ascent of warm air along the cold front may cause some severe weather, including thunderstorms or even tornadoes, to occur. With the passage of the cold front, winds will veer sharply to the northwest, temperatures will plummet, the barometer will rise sharply, and northwest winds will usher in an anticyclone with its clear skies, cold air, and lower humidities. The winds will gradually abate as the center of the anticyclone approaches, and in a day or so the anticyclone will drift on by the Lakes—and the stage will be set for arrival of the next low-pressure area from the west.

To summarize: With wind south of east, barometer falling, and all other weather clues indicating the approach of a cyclone, the cyclone is approaching from the west of northwest and will pass to the north of the observer. With wind north of east, the cyclone is most likely approaching from a point on the compass south of due west and will pass to the south of the observer. The observer should then keep in mind the weather sequences which are most probable. Although many storms are far from typical in their behavior, at least some of the features which accompany the model storms will probably be experienced. By watching for telltale shifts in the wind, the observer can then project what is in store weatherwise during the next couple of days.

5: Localized Storms

THE GREAT LAKES ALSO EXPERIENCE STORMS WHICH are more localized in their extent. Occasionally these may reach severe levels and be accompanied by high winds, damaging hail, funnel clouds, or tornadoes. Severe local storms are most likely to occur during the warm season, although in the southern portions of the Lakes they may occasionally occur in the winter. The thunderstorm is the common denominator for the severe weather effects mentioned, and it is necessary to know something about thunderstorms in general before examining their occurrence in the Great Lakes area.

Thunderstorms require warm, moist air for their development and some type of mechanism to lift the air rapidly. The requirement of warm, moist air provides a partial basis for determining their global and national distribution. Thunderstorms are much more common in the tropics than in the polar regions. In the United States, thunderstorms are most frequent along the Gulf coast and in Florida. Thunderstorm frequency drops off toward the north in the eastern United States. The Great Lakes area, however, experiences a fair amount of thunderstorm activity (figure 20). More thunderstorms occur in the southern portions than in the northern portions.

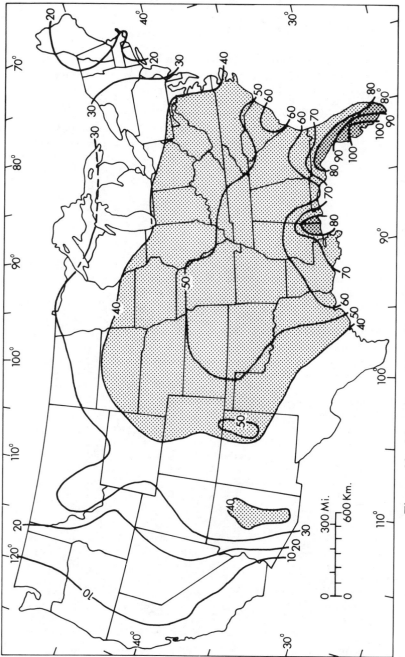

Figure 20 Mean Annual Number of Days with Thunderstorms

The requirement that the warm air be lifted rapidly in the atmosphere further defines the global and national areas of maximum thunderstorm occurrence. Lifting processes may take the form of intense heating of the surface which will start the air rising, or they may result from atmospheric disturbances such as a front. In either case, the lifting must proceed high

Aftermath of 2 April 1977 tornado at Augusta, Michigan.

enough into the atmosphere so that the thunderstorm can obtain large vertical dimensions. The tops of thunderstorms may at times reach as high as 50,000 feet (15,200 meters) or more, sustained by strong updrafts of air. If a favorable atmospheric condition is not present for lifting air to high levels, thunderstorms will not form. Along the west coast of the United States, for example, thunderstorms do not often occur except in mountain areas. Although air may be warm and occasionally moist, the downward motion of the subtropical high in the North Pacific places a "lid" on lifting.

The thunderstorm itself is a means, via the lightning stroke, by which an electrical charge differential between the earth's surface and the upper atmosphere is maintained. Thousands of thunderstorms occur each day throughout the world, and this is fortunate; for, in addition to maintaining the charge differential the thunderstorm can bring very generous quantities of rain—in fact, in some areas the major portion of summer rainfall occurs as a result of thunderstorm activity. This is true for the Great Lakes region. Occasionally, however, thunderstorms reach severe levels, and their destructive effects more than offset the beneficial rains. For the purposes of discussing the weather and climate of the Great Lakes area, it is pertinent to note that thunderstorms *do* occur in the region, that they sometimes become *severe,* and that the presence of the Lakes themselves affects the frequency of thunderstorm occurrence. We will see how in a later chapter.

Tornadoes

While they are not frequent visitors to the Great Lakes area, tornadoes are certainly not unknown. These violent localized storms are associated with severe thunderstorm outbreaks. The tornado funnel, a rapidly rotating cloud, usually blackish in appearance and in contact with the ground, is a dreaded weather feature—and with good reason. Tornadoes kill almost 100

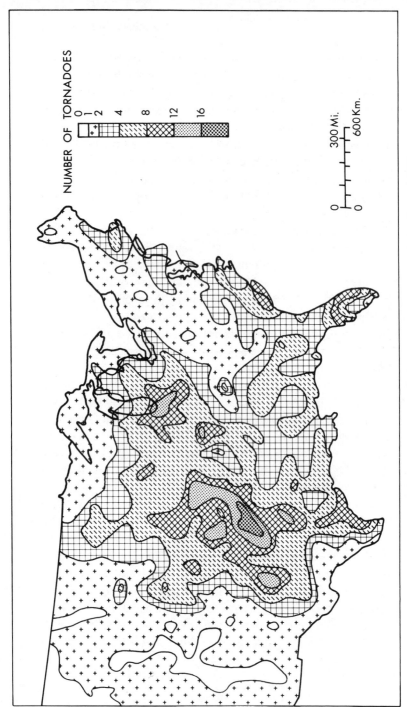

Figure 21 Mean Annual Incidence of Tornadoes per 10,000 Square Miles

NUMBER OF TORNADOES

0
1
2
4
8
12
16

300 Mi.

600 Km.

0

0

persons annually and cause damages costing millions of dollars. The two most recent major outbreaks of tornadoes—the Palm Sunday outbreak of 11 April 1965 and the 3-4 April super outbreak of 1974—each affected the southern portion of the Great Lakes basin.

Maps showing tornado frequencies in the United States (figure 21) show that the Great Lakes area is at the northeastern end of the belt of high tornado frequency that extends from north central Texas into northern Illinois. The classic *tornado alley* is located in Oklahoma and Kansas within the core of this belt. Within the Great Lakes region, tornadoes more often occur in the southwest over Illinois, Indiana, and southwestern Michigan, and frequencies decrease to the north and east. Thus residents in the southwest part of the Great Lakes basin are more likely to experience a tornado than those residing elsewhere within the basin.

Tornadoes, like the thunderstorms which spawn them, are generally warm-season phenomena. The basic requirements for their formation are similar to those for thunderstorm development—warm, moist air and a means to lift the air rapidly. However, some special additional requirements seem necessary for tornado formation. Tornadoes seem to prefer a "sandwich" type of atmosphere—that is, an atmosphere consisting of a surface layer of very warm, moist air, usually with winds from the south or southwest, and an upper layer which is colder and quite dry. The winds in the upper levels may blow from the west or northwest. Such a condition was very evident in Michigan just prior to the occurrence of the 1965 Palm Sunday outbreak. Low clouds could be seen scudding from the south. The breaks in the clouds revealed a higher layer advancing from the northwest. An hour or so later, tornadoes were occurring nearby.

With the sandwich type of atmosphere, an approaching cold front may provide the lifting necessary to set off the violent convective activity necessary for tornado funnels to form. A

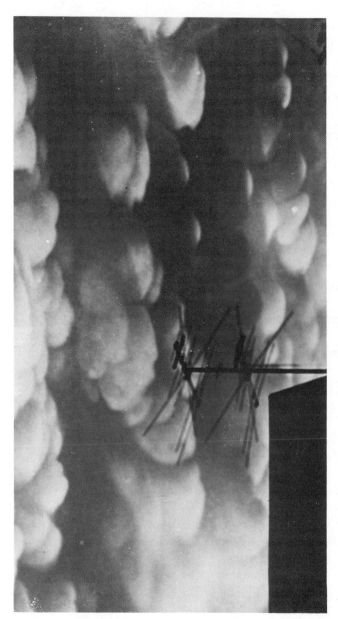

Mammatocumulus clouds, associated with severe weather, often precede tornados.

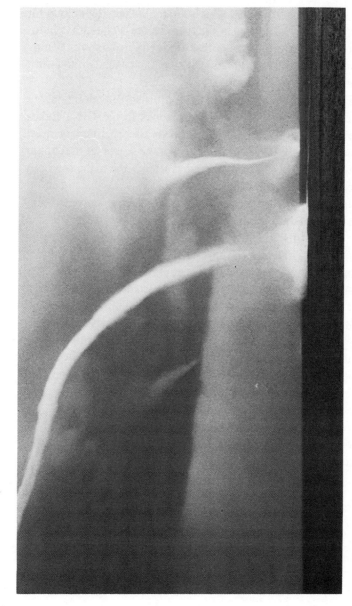

Waterspouts near the Bahama Islands. These severe weather features may also occur over the Great Lakes, especially during the autumn when the Lakes are relatively warm.

severe thunderstorm cloud which rotates is the parent cloud for the tornado. While funnels themselves are frequently photographed, the parent cloud is less frequently observed. It may extend to heights of 50,000–60,000 feet, and usually rotates in a counterclockwise direction.

The special conditions needed for tornado development are reflected in the climatology of tornadoes in the Great Lakes region. Although moist air and convective activity is most frequent during midsummer, that is not the season of maximum tornado occurrence. Tornadoes most often occur in the southern Lakes area during April, and during May in the north. In addition, while thunderstorm and convective activity seem most frequent during middle or late afternoon (when the surface is heated to its greatest extent and consequent convective activity is at a maximum), most tornadoes occur later—usually around 6:00 P.M. They can occur anytime, however.

And while most tornadoes are associated with warm, moist air at the surface, the Great Lakes area is sometimes affected by *cold-air funnel clouds*. These are funnels which develop in cold air when the more usual conditions for their formation are lacking. Such cold-air funnels may result partly from the influence of the Lakes themselves. On the other hand, the Lakes may offer some protection from the more normal type of tornado occurrence.

We will now investigate the various ways by which the presence of the Lakes may modify the atmospheric controls described in this section.

PART TWO

The Lake Effects

6: How the Great Lakes Affect Weather

THROUGHOUT THE GREAT LAKES AREA, THE LAKES themselves modify the weather and climate. At times these modifications are easily perceptible. At other times they are more subtle, discernible only to the trained meteorologist aided by special networks of instruments. Each part of the Great Lakes area experiences its own individual mélange of lake-caused weather, although certain effects are apparent everywhere.

Perceptions of the weather role of the Lakes vary. To some, the Lakes are seen as benign benefactors, mitigating the effects of otherwise harmful weather. Such may be the view of the fruit farmer in southwestern Michigan, where Lake Michigan prevents the occurrence of killing frosts in late spring and early fall. By others, the Lakes are seen in a different light—as increasing the ferocity of winds and storms, lengthening the already long and frigid winters, and obscuring the sky for long periods of time with low-hanging clouds and showers of snow.

For urban residents of Buffalo, New York, the blinding snow squalls which generate over Lake Erie are common and costly nuisances. They inundate the metropolitan area with tons of snow; expenditures for snow removal soar, and the costs to businesses are substantial. But in the hills south of Buffalo, ski resort operators rejoice—snowfall from Lake Erie means a long, profitable season.

Along the Chicago lakefront, the presence of Lake Michigan means cooling breezes during the summer, a natural, energy-saving, air-conditioning system. But in Chicago's early days, strong east and northeast winds from Lake Michigan piled up water as far inland as Wabash Avenue, and even today gales from the lake leave their destructive marks along the lakefront.

Inhabitants of Escanaba, Michigan, and Toronto and Thunder Bay, Ontario, have their air-conditioning systems too. Oppressively warm and muggy south winds in summer, which may be so enervating in other portions of the Great Lakes region, are cooled by broad expanses of water before reaching these cities. However, their spring is very late in arriving, as the balmy south breezes are chilled by the ice-clogged Lakes.

In South Bend, Indiana, and Muskegon, Michigan, the Lakes mean clouds and snow in the winter, and the Copper Country of Michigan is renowned for its heavy snowfall, much of it caused by Lake Superior. During summer along the Lake Superior shoreline, heavy fog often appears. The Great Lakes may offer some protection from tornadoes; but the tornado's less dangerous cousin, the waterspout, is common during late summer and fall, and hail may batter fruit crops. Both waterspouts and hail may be related to the presence of the Lakes.

The numerous weather and climate features which primarily result from modifications imposed by the Great Lakes are collectively called *lake effects*. In spite of their varied nature, it is possible to trace all of them to some basic physical processes occurring within the boundary layer of the atmosphere (the lower zone in contact with the lake surface). These processes occur year after year with a regularity predictable within certain limits. They stem in turn from the contrasting physical properties of land and water, which ultimately result in differences in the surface temperatures of the Lakes and the surrounding land. From these temperature differences, and the fact that the Lakes expose an all-water surface or *interface* to the atmosphere, arise the entire spectrum of lake effects.

Why the Great Lakes May Be Warmer or Colder than the Surrounding Land

Nearly everyone is aware that large water bodies such as oceans or the Great Lakes may at times have quite different temperatures than the land which surrounds them. Swimming in any of the Great Lakes in early June is usually a chilling experience, although air temperatures along the shore may well extend above 80°F (20°C). Ice floes may occasionally (as, for example, during the spring of 1972) be reported in Lake Superior as late as June, and shore ice lingers tenaciously long after spring foliage has appeared. On the other hand, campers, hikers, and fishermen in the autumn note that the waters of the Great Lakes seem tepid long after the first cold spell from Canada has whitened the land with frost. And ice does not usually begin to form on the Great Lakes until sometime after the Christmas decorations have been put away for another year.

Temperatures of the Great Lakes surface waters appear to lag behind those of the surrounding shores and never quite catch up, becoming neither as cold nor as warm as the land surface. Why is this the case? What differences exist between the Lakes and the land to explain these temperature contrasts? And how may weather and climate be affected?

How the earth's surface is heated

Let's review the ways by which the surface of the earth is heated (figures 22, 23). No matter whether on land or water, heat is gained by essentially similar processes. Surfaces become warmer if incoming solar radiation exceeds outgoing terrestrial

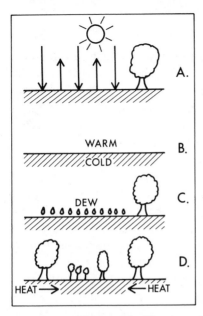

A. Incoming solar radiation exceeds outgoing terrestrial radiation. **B.** The air over the surface is warmer than the surface itself. **C.** Condensation occurs on the surface. **D.** Heat flows into the surface (insignificant on land, but possibly significant for some portions of the Great Lakes).

Figure 22   Heating of the Earth's Surface

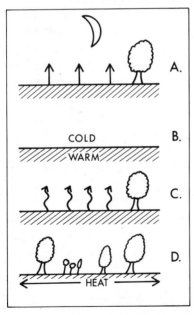

A. Outgoing terrestrial radiation exceeds incoming solar radiation. **B.** The air over the surface is colder than the surface itself. **C.** Evaporation occurs at the surface. **D.** Heat flows away from the surface.

Figure 23   Cooling of the Earth's Surface

radiation, thus creating an energy surplus which may function to raise the temperature.* Energy surpluses are normally available from about February through November in the southern Great Lakes and from March through October in the north. During the remainder of the year, outgoing radiation exceeds incoming radiation, and energy deficits exist.

If radiation exchange was the only way in which surfaces could gain heat, we would expect them to warm when energy surpluses exist and cool when deficits exist. But such is not always the case, for surfaces in the Great Lakes area may begin to warm later and cool earlier than the mean dates of occurrences of energy surpluses or deficits. Thus they must gain or lose heat in other ways.

Surfaces may also be warmed if the air above them is warmer and may be cooled if the air is cooler. Heat is transferred under these conditions by conduction and convection. The relative warmth or coldness of the air may have been acquired elsewhere—perhaps from the warm Gulf of Mexico or the Arctic regions of Canada—and moved or *advected* into the area. Thus during late winter, energy surpluses may exist over the Great Lakes region but the surfaces warm very slowly because heat is lost by conduction in warming the overlying air. And during late summer, surfaces begin to cool, even though they still experience an energy surplus, because they lose heat to the cooler air.

Heat may also be removed or added from the earth's surface in yet other ways. For example, if evaporation occurs the surface is cooled because energy is required to change the liquid to vapor (while condensation on the surface may cause it to warm). Large amounts of energy may be utilized for evaporation over water bodies, thus slowing the warm-up process. Also, in the case of a water body, an inflow of warm water may

*This energy surplus (net radiation) may also effect photosynthesis, evaporation, and melting of snow and ice.

cause the temperature to rise through *mixing*. This may be of some significance in localized near-shore areas but is only a minor factor for the Great Lakes as a whole. The land areas, of course, are immobile, so for them this process can be ignored.

To be sure, some differences exist in the amounts of energy added to the surface of the Great Lakes as compared to the surrounding land. However, those differences are not nearly large enough to account for the temperature contrasts which sometimes exist. How, then, can they be accounted for? The answer must lie in the differing ways that water and land respond to the addition of energy (figure 24).

In early April a seaman attached to the Coast Guard icebreaker
Mackinaw **finds the ice in Green Bay to exceed**
three feet in depth.

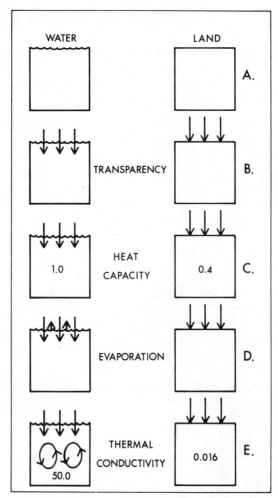

A. Equal volumes of water and land receiving equal amounts of energy *respond differently* because— **B.** Water is transparent. Radiation that is absorbed mostly at the surface of the land may penetrate some distance into water. **C.** Heat capacities differ. More calories are required to raise the temperature of water 1°C. **D.** Some energy is used to evaporate the water. **E.** Most important, thermal conductivities differ. Heat is dispersed downward in water due to its motion.

Figure 24 Differences in the Heating Rates of Land versus Water

Why land surfaces warm more rapidly than water surfaces

A physical property possessed by all substances is called *heat capacity*. Heat capacity is the amount of heat, in calories, required to raise the temperature of one cubic centimeter of a substance by 1°C. It is therefore related to the rate at which substances will warm. The heat capacity of water is 1.0 (meaning that one calorie is needed to raise the temperature of a cubic centimeter of water 1°C). However, the heat capacity of dry sand is 0.3, and that of wet soil is 0.4 (Petterssen, 1969, p. 63). This indicates that more heat is required to raise the temperature of a volume of water 1°C than to raise the temperature of an equal volume of land 1°C. Thus land should heat more rapidly than water. This fact in itself may be sufficient to account for some of the observed difference between the lake and land temperatures; but it would still not account for the large temperature contrasts observed at certain times of the year.

What else, then, must be considered? Water, unlike land, possesses a certain degree of *transparency* to incoming sunlight. Solar energy over land is mostly absorbed right at the surface; over water, a portion of the energy penetrates and disperses downward instead of being confined to the surface. Also, a portion of the energy received by a water body is utilized in the evaporation process, thus cooling the surface and retarding its increase in temperature.

All of these processes, however, are still not sufficient to account for the large temperature differences which may occasionally be observed. We need to look at yet another physical property called *thermal conductivity*. The thermal conductivity of a substance is defined as the rate of penetration of heat from the surface into the substance—or, in the case of land versus water, the flux of heat downward from the surface. Thermal conductivity is expressed in calories per centimeter per second.

Spring on the Great Lakes. The U.S. Coast Guard icebreaker *Mackinaw* performs her workaday chores.

If the water is absolutely still and the land consists of dry sand, thermal conductivities are 0.0015 and 0.0013 respectively, and heat is transferred by molecular processes. In other words, little difference appears in the downward flux of heat for still water and dry sand. However, the water in the Great Lakes is *not* still, but is constantly in motion, responding to stresses induced by winds, currents, and density differences. The thermal conductivity of water *in motion* increases to approximately 50.0, and the heat transfer occurs by eddy motion. Thus the heat flux downward within water due to mixing and turbulence is thousands of times that within land. While mixing decreases with depth, in very large bodies such as oceans it may extend to a depth of several thousand feet.

Mixing, resulting from the fact that water is in motion, primarily accounts for the much slower response of the Great Lakes to energy inflow. Were it not for mixing, the Lakes would warm at about the same rate as the land. But the constant motion of the waters in the Great Lakes distributes heat downward with great efficiency, preventing the surface from warming rapidly. Thus the surface temperatures of the Lakes lag behind those of the land during the spring warm-up.

Why the Lakes cool more slowly than the land

Another very important property of water that should be discussed at this point is that its *density* changes with its temperature. Water is most dense at 4° C (39.2° F), not at its freezing point. As the temperature departs from 4° C, water decreases in density. If water on the surface becomes more dense than that below, an overturning occurs. The surface water sinks and is replaced by lighter subsurface water. This means that as the surface waters of the Lakes cool, they become more dense and are immediatelyreplaced by warmer waters from below. Owing to this process and to the enormous amount of heat which has been stored within the Lakes, the Lakes will cool very slowly in the fall. Also, in the spring, as the temperature of the surface water rises from near freezing to 4° C (39.2° F), its density will increase, it will sink, and colder waters will rise to the surface. This overturning contributes to the slow warming of the Lakes compared to the land.

The Yearly Progression of Temperature of Lake Surface Waters

Keeping all of these things in mind, let us examine the seasonal change of mean surface temperature on one of the Great Lakes during an average year. We will use Lake Michigan for our example. As noted in a previous chapter, Lake Michigan is neither the largest nor smallest of the Great Lakes, nor is it the shallowest or deepest. The volume of water which it contains is neither the largest nor smallest, and it is neither the most northerly nor southerly of the Lakes. The progression of temperature occurring during the year in Lake Michigan is duplicated to some extent in the other Lakes, although some differences must exist because of contrasts in depth, volume, size, and latitudinal extent.

We will begin our investigation in late March at the onset of the spring warm-up. Frigid temperatures and deep snows still enshroud the northern parts of the Great Lakes basin, but the snow has melted and spring is approaching in the south. Ice lingers, however, along the shoreline of Lake Michigan and extends, in some sections, a considerable distance from the shore, depending upon the winds and the severity of the preceding winter. Most of the lake, however, is free of ice, but the water is at or near its lowest yearly temperature. Obviously, only polar bears would consider swimming in the lake at this time of year, although in only a few months the beaches will be crowded with bathers. What happens between the days of late winter, when one would perish within minutes in the icy waters, and the dog days of August, when the lake feels only refreshingly cool?

By March, as the days become longer and the sun's rays more direct, an energy surplus exists at the surface over much of the lake. The lake fails to warm much, however, as it is still losing heat by conduction to the cold atmosphere and much of the energy surplus goes to melt the ice. However, by late March in the south and early April in the north, the air has become warm enough so that it warms the lake, and the energy surplus also becomes larger.

The lake then begins to warm, but very slowly. As we have seen, much more heat is required to warm the water surface than the surrounding land chiefly because of the great thermal conductivity of the lake. Additionally, as the surface waters of Lake Michigan slowly warm from near 0°C (32°F), they become more dense. They then sink and are replaced by colder, less dense water from below. With this continual overturning and replacement, heating goes on very slowly until the entire water column achieves a temperature slightly above maximum density (4°C or 39.2°F). Where the lake is deep, a considerable amount of time elapses before the entire column is warmed to maximum density. In shallow places, this point is reached more quickly. Maximum density of a water column in the middle of the lake may occur anytime between the second week of May and the second week of June, depending on how cold the lake was at the end of the winter and how warm the spring months were. By this time, of course, late spring and early summer hot spells have warmed the surrounding shores, and the difference between the lake and land temperature is large. No wonder the beaches are dotted with sunbathers in early June—but only a few hardy swimmers brave the chilly waters.

Once the lake has warmed to maximum density, it is poised for a more rapid warm-up. As the surface temperatures rise above 4°C (39.2°F), the water expands and the density decreases. The warmed surface water then literally floats on the colder, denser water beneath. A stratification takes place as the surface waters continue to warm, and from June to mid-July the temperatures increase rapidly. Mixing by winds and currents

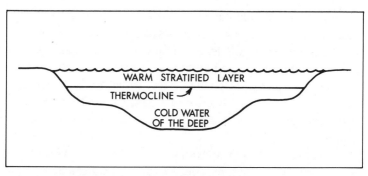

Figure 25   Midsummer Conditions in the Lake

deepens the stratified layer as the season progresses, extending it to 40 or 50 feet beneath the surface by midsummer (figure 25). Below lurk the cold denser waters of the deep, which change little in temperature throughout the year. The boundary zone between the two layers is called the *thermocline,* and it marks a rather abrupt transition between the warmer surface waters and the cold bottom waters. The depth of the thermocline may vary considerably with the weather conditions, sometimes rising close to the surface, at other times descending to greater depths.

The irregular bottom contours of the lake modify the warm-up process somewhat. The lake waters are very shallow near the shore and much deeper near the center. Consequently the shallow waters near shore heat up much more rapidly than the deeper waters in mid-lake because the downward dispersal of heat by mixing is limited by depth. As the inshore waters exceed 4°C in temperature, become stratified, and warm further, the waters in the deep lake may not yet have reached 4°C. A boundary zone called the *thermal bar* (figure 26), consisting of water at or near maximum density, forms between them and extends down into the lake. The thermal bar moves progressively farther and farther out into the lake as the warm-up continues and eventually disappears as all of the lake water reaches greatest density. The thermal bar is important because it acts as a barrier which may prevent the mixing of

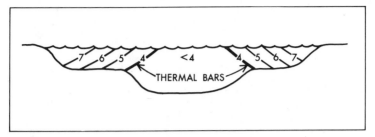

Figure 26   Thermal Bars (in degrees Celsius)

inshore waters with those farther out in the lake, thus trapping pollutants along the shore.

Sometime around mid-August or early September the lake begins to cool. As the surface water chills, it becomes more dense, sinks, and is replaced by warmer waters from the depths. These, in turn, are cooled. Mixing by storms and currents extends the cooling throughout the entire upper stratified layer until eventually the whole lake is cooled to the temperature of maximum density, 4°C. Thus the cooling process, like the heating process, also goes on very slowly, and a large temperature contrast between the lake and the land is again generated by late fall and early winter. But at this time of year, the lake is much warmer than the land. Once the lake has cooled to maximum density (not until sometime around the first of January), the surface layers may cool further; and a sort of reverse stratification may develop, with the colder surface waters floating on top of the slightly warmer and denser deep waters.

Stable season *versus* unstable season

Figure 27 illustrates the mean surface water temperatures near the center of Lake Michigan at various months of the year in comparison to the mean temperatures of the surrounding land. Note that two periods exist, one about mid-March and another in early or middle August, when the lake temperatures

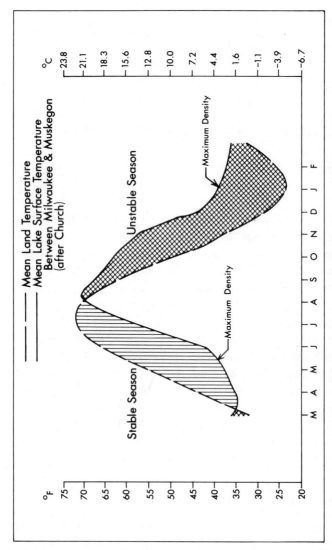

Figure 27 Mean Land Temperatures versus Mean Surface Lake Temperatures in Southern Lake Michigan Area

are about equal to the shore temperatures. From mid-March to August, the lake is colder than the land with the greatest difference occurring from mid-May to early June. During the fall cooling season, the lake remains warmer than the land from about late August until middle or late March, with the largest difference occurring in late November and early December.

The period when the lake is cooler than the land is known as the *stable season,* while its opposite, the *unstable season,* occurs when the lake is warmer than the land. All of the Great Lakes have stable and unstable seasons, with differences in the time of onset, duration, and intensity caused by contrast in lake size, depth, and latitudinal extent. While Lake Huron behaves similarly to Lake Michigan because depth, size, and latitude are not greatly different, Lake Superior is larger, deeper, and farther north than Lake Michigan. Consequently the warming season does not begin until later in the spring, and surface temperatures never attain the high values of the lower Lakes (figure 28). On the other hand, Lake Erie, being smaller and quite shallow, warms rather rapidly in the spring and cools more rapidly in the fall. The peaks of the stable and unstable seasons occur earlier in the spring and fall.

Another interesting fact concerning Lake Erie is that in spite of its southerly location, an extensive ice cover is commonly formed while the other Lakes may be relatively free of ice (figure 29). This is because of Lake Erie's shallowness, which permits more rapid fall and early winter cooling. After it is covered with ice, it behaves more like the land surface. Lake Ontario, although not greatly different in latitude, is a rather deep lake, so extensive ice rarely forms. The presence or absence of ice on the Great Lakes has some very important implications to the climate of surrounding areas, as will be shown later.

The occurrence of a large temperature difference between the Lakes and surrounding land is the key to the weather modification properties of each lake. The larger the temperature

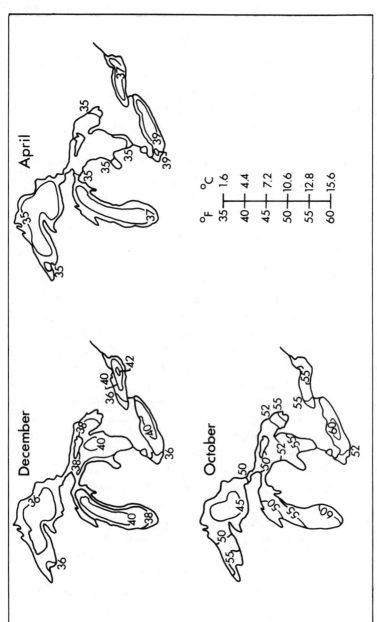

Figure 28   Mean Great Lakes Water Surface Temperatures for December, October, and April

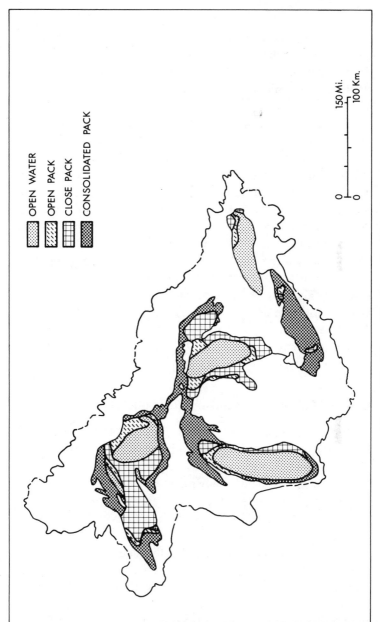

Figure 29 Mean Maximum Ice Extent

OPEN WATER
OPEN PACK
CLOSE PACK
CONSOLIDATED PACK

150 Mi.
100 Km.

0
0

difference, the more marked are the various features of weather and climate which the Lakes may cause. Thus the influence of the Great Lakes should be most marked at the peaks of the stable and unstable seasons, when the temperature differences are sharpest.

Obviously, at these times the lake surface temperatures are not uniform everywhere. Sharp changes can occur within short distances and from day to day. The shallow shore areas are much warmer during the stable season but become colder and ice-clogged during the unstable season. In the summer, onshore breezes may pile up warm surface water along the beaches, making swimming a delight. An overnight change to an offshore wind transports the warm surface water offshore and allows cold water from the depths to replace it, resulting in a 10°–15° drop in temperature overnight. This process is called *upwelling* and may be a common and unwelcome experience to bathers along the upwind shores (generally the west shores of the Lakes). But in spite of these local variations, the mean surface temperatures of the Great Lakes exhibit the established seasonal pattern of change described.

How the Lakes affect wind velocities and direction

We need yet consider one other difference between the Lakes and the surrounding land which may affect meteorological processes: that the surfaces of the Lakes are relatively smooth, while the land area is rough. To be sure, high waves sometimes occur on the Lakes, but they are small in comparison to the hills, valleys, and rough surface features of the surrounding land.

The smoother lake surface means that there is less friction to slow the movement of air—hence winds are stronger over the lake and along the shore. When winds move from the rough land to the smooth lake, they increase in velocity and develop a slight downward motion. On the other hand, when the winds

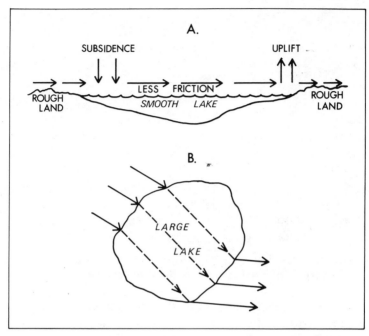

A. Velocity increases over the lake because of reduced friction. Subsidence occurs on the upwind shore as wind speed increases; uplift on the downwind shore as wind speed decreases. **B.** Direction of the wind changes slightly as air flows over the lake. Lessened friction and higher velocity increases the coriolis effect.

Figure 30   Changes in Wind Direction and Velocity as Air Flows across a Large Lake

encounter rough land on the downwind side of the Lakes, they slow down and a slight piling up accompanied by an upward motion may occur. In addition, the change of speed as winds cross from one type of surface to another causes a small change in wind direction (figure 30). This is because the coriolis effect, as described previously, is partly a function of velocity. The change of wind direction can also contribute to upward or downward motions along the shores.

Keweenaw Bay, Lake Superior, March 1976. The build-up of ice restricts interaction between the lake and atmosphere.

How the Lakes Affect Weather
during the Stable and Unstable Seasons

It is important to note that the Great Lakes affect weather and climate in contrasting ways during the stable and unstable seasons. During the stable season when the Lakes are cold relative to the land, winds crossing the Lakes are chilled and cause cooler temperatures on the surrounding land. Evaporation rates are low, and the transfer of moisture from the Lakes to the air is minimized. Hence the tendency of this air to rise is lessened. The cooling of the air also causes localized high-pressure cells to form over each lake. These features may not be discernible on the weather maps produced by the National Weather Service because the spacing between weather stations is too great, but they become apparent when special networks of instruments are employed. As the Lakes cool the air, the temperature near the surface becomes colder than that aloft, forming an *inversion* (that is, instead of the normal fall of

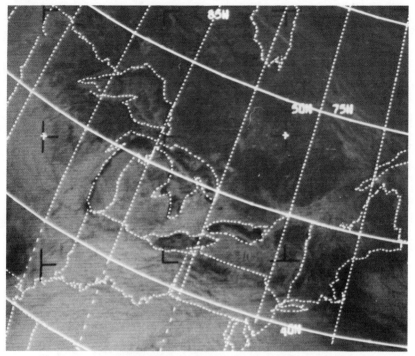

ESSA-9 satellite photo showing cloud deck over Lake Michigan due to overpassing cold air. Note the snowband over Lake Ontario.

temperature with increasing height in the atmosphere, temperatures may actually rise). The inversion suppresses upward motion and may have some important implications for air pollution in the Great Lakes region. Cyclones which approach the area during the stable season tend to weaken because the Lakes fail to supply heat. On the other hand, anticyclones become stronger. The stable season, then, is a period when the Lakes exert a steadying influence on the air, suppressing its tendencies to warm rapidly and to rise.

The unstable season ushers in a reversal of all of these processes. Now the Lakes are warmer than the land. Winds passing over the Lakes acquire heat and, as evaporation is rapid, much moisture. As the air is warmed, it becomes buoyant and rises. Inversions, if they exist, are destroyed or "lifted" to greater heights. Warmth is carried to the surrounding land, and the heat and moisture acquired by the Lakes may form clouds. Cyclones approaching the Lakes are intensified due to the acquisition of heat from the Lakes. Anticyclones, however, rarely pass directly over the Great Lakes, but are diverted west and south or slip north of the Lakes into eastern Canada, where they are known as *glancing highs*. The air passing over the Lakes becomes restless and turbulent, and being warmed, has a tendency to rise.

Thus we have seen that there are many ways by which the interface of the Great Lakes interacts with the boundary layer of the atmosphere. But what are the specific weather features which result from these interactions? And how do they affect the residents of the area? Can man achieve any control over the lake-caused weather features? Are such controls desirable? We will attempt to answer these and other questions in the following chapters.

7: The Fogs

NEAR THE TIP OF MICHIGAN'S KEWEENAW PENINSULA, a spiny ridge called Brockway Mountain rises nearly 700 feet above Lake Superior. From the summit on many summer days, banks of fog and mist can be spotted far out over Lake Superior. Satellite photographs clearly depict these grayish masses hovering over the Great Lakes. From Chicago to Duluth to Buffalo to Toronto, Great Lakes cities must occasionally cope with a chilly, wet blanket of fog which can be very persistent at times, but can also be quickly dissipated by a sudden shift of the wind. Because some basic interactions between the Lakes and the atmosphere are exemplified by the formation of fog, we will begin our discussion with this lake-caused feature.

What Is Fog?

Fog occurs when very small water droplets (or ice crystals) are present in the air near the earth's surface so that visibility is reduced to less than 1 kilometer (about ⅝ of a mile). The droplets are so tiny (from 1 to 50 microns* in diameter as

---

*A micron is one ten-thousandth of a centimeter.

compared to about 1,000 microns for the average sized rain-drop) that they are literally suspended, constantly evaporating and reforming, in the air. Fog has often been described as a low stratus cloud in contact with the surface—and in truth, being in a fog is very much like being in some types of clouds. In both cases, the water droplets are very small and lack the weight and volume necessary to fall out as precipitation, though processes may occur within some clouds which cause the droplets to gain weight and volume, whereby they fall from the cloud as precipitation.

In fogs, some growth of the small droplets may cause precipitation in the form of drizzle. Artificial stimulation of this growth process which, if carried far enough, may lead to dissipation of the fog is also possible. This may be desirable when restrictions of visibility caused by fog may result in transportation (particularly air transportation) tie-ups.

How fog is formed

In order to understand how fogs occur and why they are common to the Great Lakes, we must know something about the role of moisture in the atmosphere. The atmosphere, as previously stated, consists of a mixture of gases, including nitrogen, oxygen, argon, carbon dioxide, ozone, and a host of minor gases. Water in the gaseous state (water vapor) is also present in varying quantities (usually from about 1%–5%). Water vapor enters the atmosphere through evaporation, mostly from the ocean surfaces but also from lakes, rivers, soil, vegetation, and other features of the earth's land surface which contain water.

The amount of water vapor which air can hold is strictly determined by temperature. The higher the temperature of the air, the more water vapor the air can hold. At any given temperature, the air has a maximum capacity of water vapor. If this capacity is reached, the air is said to be *saturated*. This

Fog bank rolling onshore at Milwaukee, Wisconsin.

means that no more vapor can be added to the air at that temperature unless some of the vapor *changes state,* becoming a liquid or solid. This process of water vapor changing to liquid water is known as *condensation,* which is familiar to most of us. With air at the saturation point, condensation begins to occur; or, in other words, the moisture present in the vapor state begins to condense into small water droplets.* If the saturation occurs at subfreezing temperatures, the vapor may change directly to a solid (ice crystals) without passing through the intermediate liquid state. This process is known as *crystallization.*

The amount of water vapor in the air can be expressed in terms of the partial pressure exerted by the water-vapor fraction of the atmosphere. As a mixture of gases, air exerts a total pressure (measured by barometers), while each individual gas exerts a partial pressure. Oxygen, nitrogen, and water vapor each exert partial pressures. The partial pressure exerted by

---

*The water condenses around suitable *nuclei*, or tiny particles present in the atmosphere.

water vapor is called *vapor pressure* and can be expressed in the same units as total atmospheric pressure (in inches of mercury or in millibars). The maximum vapor pressure which air can hold at any given temperature is called the *saturation vapor pressure,* also usually expressed in inches of mercury or in millibars. Saturation vapor pressures are strictly a function of temperatures, and they increase as temperatures increase.

Although newspapers, radio, television weathercasts, and other popular weather information sources rarely mention either vapor pressures or saturation vapor pressures, they very commonly announce the *relative humidity,* which is given in a percentage figure and is based approximately on the ratio of vapor pressure divided by saturation vapor pressure. In other words, VP/SVP $\times$ 100 = relative humidity in percent.* One definition of relative humidity calls it an expression, given as a percentage, of the amount of water vapor actually in the air compared to the amount which the air can hold at that temperature. If the relative humidity becomes 100%, the air is saturated and can hold no more water vapor without the occurrence of condensation.

What has all this to do with fog, and especially with the fogs of the Great Lakes? It was stated that fog consists of tiny water droplets suspended in the air near the surface. These water droplets are condensation products, and they are present because the air near the surface has acquired a relative humidity of 100%.† Thus the vapor is condensing, or changing state. The tiny water droplets which result compose the foggy veil which restricts visibility. Whenever substantial masses of air near the surface become saturated, the potential for fog is present. But what causes the air to become saturated?

---

*In the notation used by meteorologists, vapor pressure is represented by e and saturation vapor pressure by e_s. Thus, $e/e_s \times 100 = $ RH.

† Actually condensation may begin in the atmosphere at relative humidities of less than 100% due to the presence of hygroscopic nuclei (substances which attract water chemically).

Some simple arithmetic will tell us that two ways exist by which the ratio VP/SVP can be changed so that it will equal unity, or so that saturation will occur. The vapor pressure could be increased by actually adding more water vapor. This could be accomplished by evaporation. The saturation vapor pressure could be decreased by cooling the air. This would decrease the capacity of the air to hold whatever moisture is already present. If the cooling could be carried far enough, saturation would be reached. It is important to remember that in the atmosphere, *cooling is the process which results in most condensation forms* (including fogs), although additions of water vapor may occasionally become important.

As a practical matter, all of these scientific principles combine to produce a very straightforward recipe for fog: (1) Take a layer of air near the surface with a high moisture content; (2) bring the air to saturation (the ratio VP/SVP × 100 will equal 100%); this can be done either by adding moisture (increasing the vapor pressure) or cooling the air (decreasing the saturation vapor pressure); in most cases, it will be more expeditious to cool the air; (3) fog will result. Can be kept well as long as air remains at saturation. Caution: Keep air cool; warming will result in increased capacity of the air to hold water vapor; consequently fog will dissipate.

Thus in order to explain the fogs of the Great Lakes, we will be examining, in most cases, the way in which masses of air near the surface are cooled substantially so that their capacity to hold their moisture is lessened and saturation (100% relative humidity) occurs.

How the Great Lakes May Cause Fog to Form

Advection fogs

We have seen that at certain times of the year, the surface temperatures of the Great Lakes may be considerably colder than the temperatures of the surrounding land. This is normally the case from about mid-March until the end of summer, with the largest temperature difference occurring sometime around mid-May or early June. This is the stable season when the Lakes act like giant refrigerators, cooling the air which moves across them and chilling the immediate shorelines. During the stable season, large masses of air moving across the Lakes are cooled and stabilized.

Suppose a mass of air carried by light winds drifts slowly from the southern portions of the United States toward the Great Lakes region during the height of the stable season. This mass of air is likely to have a rather high temperature and moisture content (an mT air mass) due to its origin in southern latitudes near a large source of moisture (the Gulf of Mexico). As the air approaches the Great Lakes, its relative humidity is probably quite high—in other words, it holds just about all the water vapor it can without condensation occurring. If the air should be suddenly cooled, condensation would occur and fog formation would be possible. For all masses of air, the relation between the air temperature and the amount of water vapor contained in the air allows a computation to be made of the *dew-point temperature,* defined as the temperature to which the air must be cooled to achieve saturation. In the case of our warm, humid air mass moving toward the Great Lakes, the dew-point temperature would be the temperature to which we must cool the air in order for condensation to begin; and a considerable thickness of air near the surface must be cooled to the dew point for a fog to form. Dew-point temperatures are determined on a regular basis and are plotted on weather maps along with air temperatures, thus giving the forecaster an indication of how close the air is to saturation.

As the warm, humid air drifts over Lake Michigan (figure 31), it is chilled by the cold lake waters, losing heat by contact with the lake surface. Studies have shown that cooling of the air occurs extremely rapidly in the lowest 150 feet (45

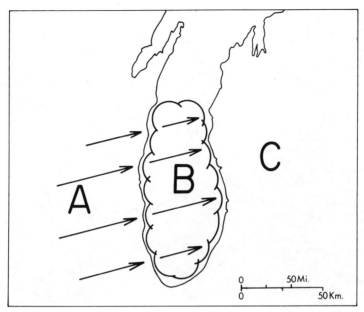

A. Warm, moist air, temperature 85°F (29.4°C). **B.** Lake temperature 50°F (10°C). The air is immediately cooled as it begins to cross the lake and fog forms. **C.** A lake breeze during the day may bring the fog a short distance onshore. However, fog generally dissipates quickly upon encountering the warmer land surface.

Figure 31   Formation of Advection Fog over Lake Michigan during the Stable Season

meters) or so within a distance of 30 miles (48 kilometers) from the shore (Bellaire 1965). Above this chilled layer, the air remains warmer, thus causing a temperature inversion. With a relatively cold lake surface temperature, as is likely to occur at the height of the stable season, it is possible that a considerable thickness of the air will be cooled below the dew point, causing the condensation of tiny water droplets which gives us fog (figure 31).

This particular type of fog is known as an *advection fog*.

Fog lingers off the southern shore of Lake Erie.

The term *advection* refers to horizontal movement—in this case the horizontal movement of warm, moist air over the much colder surface of Lake Michigan. Obviously this type of fog would not be likely to occur over the Great Lakes during the unstable season when the Lakes are warmer than the surrounding land. Therefore it is largely a spring and summer occurrence.

Breezes from the lake may bring the fog onshore for a short distance during the day. But the fog is likely to persist only along the immediate shoreline because the much warmer land temperatures which it encounters will quickly dissipate it by increasing the capacity of the air to hold water vapor. Thus picnickers and bathers may escape the fog by moving only a quarter-mile or so inland. The fog may shroud Chicago's lakefront, but the Loop may remain fog-free. The fog will persist along the lakeshore areas until a change occurs in the total weather situation. Winds may change direction so that warm, humid air is no longer advected over the lake, or a much drier air mass may replace the warm, humid air. With a drier air mass, the cooling induced by the lake may not be sufficient to lower the air temperature to its dew point. Strong winds or an increase in wind velocity may be sufficient to break up the fog, since strong winds cause turbulence which destroys the inversion and mixes the warmer air aloft with the colder air near the surface.

Winter fogs: "Arctic Sea smoke" —Another type of advection fog

Fogs also occur over the Great Lakes during the winter while the lake waters are relatively warm. These fogs result from different processes, since obviously the Lakes do not cool the air at this time of year.

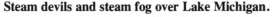

Steam devils and steam fog over Lake Michigan.

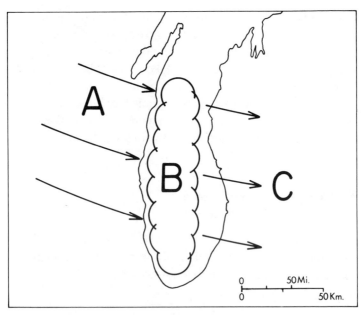

A. Cold, dry air, temperature 0° F (−17.8° C), dewpoint temperature −17°F (−27.2°C). **B.** Lake temperature 50° F (10°C). Rapid evaporation from the warm lake into the cold air saturates the air. Fog forms. **C.** Temperature on the downwind shore 10°F (−12.2°C). Dewpoint 4°F (−15.6°C). Heat and moisture are added to the air from the warm lake surface.

Figure 32   Formation of Steam Fog over Lake Michigan during the Unstable Season

Winter fog over the Lakes may occur as the result of very cold air from the Arctic moving over much warmer water. In this case, the fog does not result from the cooling of the air but is due to an actual increase in air moisture content. As cold air flows over warm water, the evaporation process occurs very rapidly; accordingly, the vapor pressure of the air rises sharply (figure 32). As with the summer fog, the net result is to bring the relative humidity to 100% so that condensation occurs. This fog is known as *steam fog* or *Arctic Sea smoke*.

Steam fog may occur when air temperatures are 5° to 40°C lower than the water and with winds from calm to gale force (Saunders, 1964). The fog depth may vary from 1 to 1,500 meters, and on occasion distinct fingers or columns of vapors called *steam devils* can be seen rising from the steam layer (Lyons and Pease, 1972).

Advection-radiation fogs

Yet another fog may occur in the Great Lakes area as a direct result of the presence of the Lakes. This fog may form in the fall or late summer when the waters of the Great Lakes are still warm but when the nights are beginning to lengthen. It chiefly occurs over the shore areas of the Great Lakes and is fairly common, forming when large parcels of air become warmed and humidified over the Great Lakes. As the air drifts slowly over the land at night, radiation processes cool the land surface rapidly, and the air in turn is chilled. The resulting fog is called an *advection-radiation fog,* combining the processes of advection and radiation. The high moisture content results from the air's residence time over the lake, but the cooling which causes the fog development occurs after it moves inland (figure 33).

Since the Lakes have ample opportunities to interact with the atmosphere to cause cooling or to add moisture, it is not surprising that the incidence of fog is rather high over, and along the shores of, the Great Lakes. On many occasions, fog may also develop independently of the presence of the Lakes—as, for example, when rains have raised the humidity of the air and nocturnal radiational cooling then causes the dew point to be reached. Additionally, many fronts are positioned in the Great Lakes region throughout the year. Fogs due to mixing of warm air with cooler air, or due to increases of water vapor in the air resulting from evaporation of raindrops, can occur along these fronts.

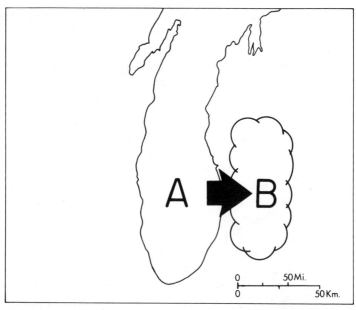

A. Water temperature 65°F (18.3°C), air temperature 72°F (22.2°C); dewpoint 66°F (18.9°C). Air over the lake acquires moisture from the lake and drifts inland. **B.** Nocturnal radiation cools the land, which in turn lowers the air temperature to the dewpoint; fog forms.

Figure 33   Advection-Radiation Fog Occurring in the Fall

Some Rules of Thumb for Planning Lakeshore Outings

Suppose you have plans for a picnic or outing along the shores of one of the Great Lakes. It is always a disappointing experience to encounter a gray pall over the shoreline while the sun may be shining a half-mile or so inland. What rules of thumb can be employed to anticipate summer fog over the lake and on the shoreline?

First, remember that these fogs tend to occur at the height of the stable season when lake temperatures are much cooler than the land temperatures. Thus May and June are prime months for fogs, although they can also occur later in the summer, particularly over Lake Superior. Second, in order for fog to form, there must be warm, moist air with light winds. Usually a light flow of air from the south or southwest is involved, and dew points should be rather high. With northerly or northwesterly winds, the air is likely to be too dry for fog development, although this is not always true. Beware of humid, sticky days with light air flow during the early summer; what appears to be a perfect day for a beach excursion may turn into a fog bath along the shore. It's best to call a Coast Guard station near your destination for visibility reports before you embark on a disappointing outing.

8: The Clouds

THERE ARE SEVERAL WAYS IN WHICH ONE MAY SEE what the earth looks like from far up. One way is to become an astronaut—for most of us, an unachievable ambition. The next best way is to view imagery of the earth from meteorological satellites or space programs such as Landsat and Skylab. One thing the meteorological satellites tell us is that a rather substantial portion of the earth's surface at any given time is covered with clouds. A primary goal of the meteorological satellite is to view this cloud cover and to convey information about its distribution and arrangement. The Landsat and Skylab programs have a different goal. Their primary purpose is to observe features at the surface of the earth, and clouds become an impediment in accomplishing this objective.

Two Landsat satellites orbit the earth in such a path that they travel over the same area once every 18 days. Over the upper Great Lakes, it was 2 years before a cloud-free photograph could be obtained. Some scientists feel that Landsat's 18-day cycle contributed to the lack of cloud-free imagery. An 18-day cycle is a multiple of a basic 6-day weather cycle in the atmosphere: This means that if cloudy weather appears on the first day, clouds will likely reappear on the sixth, twelfth and eighteenth day, and so on. This may be true, but certainly one reason it took so long to obtain cloud-free pictures is simply that the Great Lakes region is so cloudy!

Climatic statistics tell us that, indeed, during the winter, the Great Lakes are among the cloudiest areas in the United States, if not in the entire world. Data on cloud cover reveal that only some sections of the Pacific Northwest are cloudier than the Great Lakes in winter. Sun-lovers in the Great Lakes area fare much better in the summer, although the Great Lakes are far from being the most cloud-free part of the country even at that season. In the Great Lakes, seasons are defined as much by varying cloud cover as by temperature, and the winters seem longer and colder than they actually are because so little is seen of the sun. This lack of sunshine during the cold season carries with it a variety of psychological effects and even affects the physical well-being of the residents, since sunlight is needed for the production of vitamin D, a necessary vitamin component. We might also conclude that the Great Lakes are not located in an especially favorable position for the future utilization of solar energy if the basic resource is lacking during much of the year. While feasible, the use of solar energy systems would be more expensive than for sunnier parts of the country, assuming that alternative energy prices are comparable in both areas.

Clouds are commonplace in the Great Lakes region partly because the Lakes themselves are excellent cloud factories during certain times of the year. At other times, they may actually suppress clouds, and this is one reason why a large seasonal difference occurs in the amount of cloudiness.

How Clouds Are Formed

Clouds are certainly among the most vivid of weather features. They are indicators of past and ongoing atmospheric processes as well as presagers of future weather events. Clouds, like fogs, consist of very small water droplets, ice crystals, or both. Like fogs, the droplets or crystals are literally suspended in air, being of insufficient volume and mass to fall to the surface. Also, like fogs, most clouds are caused by the cooling of moist air.

Unlike fogs, however, which form at or near the surface, clouds form at a variety of heights. They may be found at very high altitudes, perhaps 30,000 or 35,000 feet (about 10,000 meters) above the surface—or even higher. Thus the method of cooling responsible for most fogs over the Great Lakes (that is, cooling resulting from the loss of heat by conduction and radiation as warm, moist air moves over a much colder surface) cannot be responsible for the formation of clouds.

Some other type of cooling process, one which effectively cools large parcels of moist air to their condensation point, must be occurring in order for cloud formation to occur. This process is called *adiabatic cooling,* and it is associated with air which ascends or is uplifted within the atmosphere.

Vertical motion in the atmosphere is not of the magnitude of horizontal motion (which we call wind). When vertical motion does occur, it usually is at a slow rate (except in cases involving severe thunderstorms, where rapid vertical motion is not unusual). But vertical motion is extremely important because *adiabatic processes* accompany vertical displacements of air.

Understanding adiabatic processes

To understand what an adiabatic process is, we must first review some basic physics. Air consists of a mixture of gases. The temperature of a gas or mixture of gases is proportional to the internal energy of the gas or to the total energy of its molecules. The *first law of thermodynamics* states that the internal energy of a gas (temperature) can be altered by either the addition or removal of heat (obviously), or by having work performed on or by the gas (not quite so obviously). An adiabatic process is one in which the internal energy of the gas is

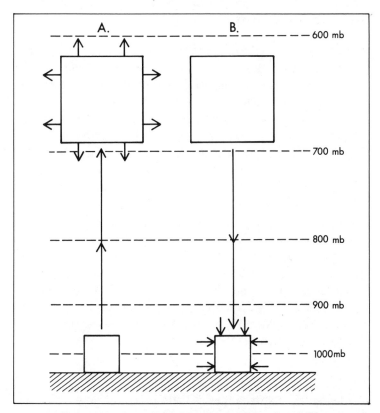

A. Lift an air parcel from the surface to a higher level. The parcel is subjected to reduced pressure, expands, and does work on the surrounding air. Internal energy is thus decreased and cooling occurs. **B.** Lower an air parcel from a height to the surface. The parcel is subjected to higher pressures, it is compressed (work is done on it), and internal energy is increased, causing warming to occur.

Figure 34   Adiabatic Processes in the Atmosphere

altered by having the gas perform work, or by having work performed on the gas. No heat is actually added to, or removed from, the gas—although the temperature of the gas changes.

To illustrate the role of adiabatic processes in the atmosphere, let us refer to figure 34. In that diagram, air which rises is subjected to reduced pressures (as the pressure of the atmosphere decreases with height). As the air rises and expands because of reduced pressure of the surrounding air, it performs work on the surrounding air which reduces its own internal energy or temperature. Conversely, when a parcel descends through the atmosphere, it is subjected to increased pressures. The parcel is compressed, and the work done on the parcel results in an increase in its own internal energy—or an increase in temperature. More simply stated, when air rises, it is cooled; when air descends, it is warmed.

The implication to cloud formation should now be obvious. When masses of air are lifted, their capacities to hold moisture are reduced because of adiabatic cooling. If the air is moist enough and the lifting sufficient, the air may be cooled to its condensation point and clouds may form. Thus any process which causes lifting to occur may potentially result in the formation of clouds. On the other hand, processes which cause downward motions (subsidence) are unfavorable for cloud formation.

Atmospheric disturbances as lifting mechanisms

Logic would suggest that since the Great Lakes seem to be quite cloudy, lifting processes must occur very frequently over the area. What mechanisms might be responsible for upward motions in the atmosphere? We have already noted that the jet stream is frequently located aloft over the Great Lakes region in association with a cyclone at the surface. The jet may provide the impetus for a lifting motion near the surface by causing divergence aloft. To compensate for the divergence aloft, air currents converge near the surface and rise to higher levels (figure 35). Divergence (and convergence) zones develop along the flanks of jet streams as they flow along at high levels and may cause cyclones, which are vast systems of rising air. As mentioned previously, few parts of the country have such a high

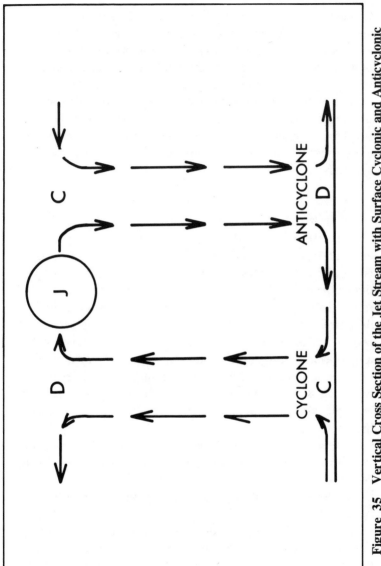

Figure 35 Vertical Cross Section of the Jet Stream with Surface Cyclonic and Anticyclonic Associations (*D* = divergence; *C* = convergence)

frequency of cyclones as the Great Lakes. With their jet stream associations at the higher levels, the lifting action of frequent cyclones which travel through the Lakes area is the major cause of the large amount of cloudiness experienced. Remember also that accompanying the cyclones are fronts, caused when air masses of different temperatures converge. Lifting also occurs in frontal zones. Both cyclonic and frontal activity is most frequent and intense during the winter, and this is an important reason why Great Lakes winters are much cloudier than the summers.

Orography

Air can also be lifted in other ways. Moist air currents that encounter mountain ranges are forced to rise and consequently cool adiabatically. This is the reason why the windward slopes of mountains are often quite cloudy and receive heavy precipitation. Lifting caused by mountains or rough terrain is called *orographic lifting*. No high mountains exist in the Great Lakes region, and orographic lifting is not of major importance in the area. Occasionally in the winter, however, orographic lifting may cause clouds to form. The air may have a dew-point temperature only a few degrees below the air temperature. For example, assume an air mass is moving across the Great Lakes area with a temperature of 31°F (−0.5°C) and a dew point of 29°F (−1.6°C). Only a slight amount of cooling would be necessary for the air to begin condensing—it would need to be cooled only 2°F.

Up to this point, the *rate* at which air cools as it rises has not been mentioned. We have said only that it cools. To be more specific, air which rises and cools adiabatically does so at a constant rate until a cloud begins to form. This constant rate is known as the *dry adiabatic rate* and has a value of 5.5°F per 1,000 feet (about 3°C per 300 meters). This means that all parcels of air which are lifted within the atmosphere cool 5.5°F for each 1,000 feet of lifting until the point is reached where the air is saturated and a cloud forms. (Conversely, subsiding air *warms* at the rate of 5.5° per 1,000 feet.) In the example cited, our moist, chilly air mass would need only a small amount of uplift (considerably less than 1,000 feet*) before it would begin to condense.

Convection

Another way by which air is caused to rise that we should consider is the lifting which results when the sun strongly heats the surface of the earth. This is called *convective lifting* and is most important during the warm season. The *thermals* or updrafts which result when portions of the earth's surface are strongly heated (as the air becomes less dense and rises) may be responsible for the prevalence of white, puffy *cumulus* types of clouds in the summer. In the Great Lakes area, this type of lifting is not of major importance in the winter because the land remains very cold. During summer, it may play a major role in cloud formation but is probably less important than atmosphere disturbances.

Clouds and the Great Lakes

How, then, do the Great Lakes play a role in forming clouds? Cold air which crosses any of the Great Lakes in winter will be warmed as the lake is traversed. In addition, moisture will be added to the air by evaporation from the lake surface. Most often during the winter, the air which is advected across the Lakes is colder than the water. The processes of heating the air and adding moisture are perfect inducements for cloud formation. Added moisture raises the dew point of the air and

*Condensation would not occur at 29°F, the surface dew point, but at a value slightly less than that, because as the air expands, the water vapor has a greater volume to saturate.

Cloud lines over Lake Huron photographed during a research flight sponsored by NCAR (The National Center for Atmospheric Research).

brings it closer to saturation; warming of the air causes it to rise, just as air rises over the heated land surface during the summer; add some high hills on the downwind side of the Lakes for orographic lifting, and we have a perfect cloud-generating machine which continues to run as long as all its parts are functioning properly.

Consequently the winters in the Great Lakes area tend to be extremely cloudy for two reasons. First, the frequency of fronts and cyclones in the area is high, and each of these features is a lifting mechanism of great effectiveness. Second, the heat and moisture added by the Lakes causes air to rise and condense, so that the cloudiness generated by the Great Lakes is added to the cloudiness provided by cyclones and fronts. No wonder that one seldom sees the sun for long periods of time!

During the stable season when the Lakes are cooler than the land, the Lakes act to suppress certain types of clouds. At this time (late March to September), air which crosses the Great

LANDSAT I photograph showing dissipation of cumuli at the western end of Lake Ontario during stable northerly flow.

Lakes is usually cooled—which means that instead of tending to rise, it has just the opposite tendency, and clouds which already exist may be dissipated. Also, the cyclone-front family takes up its summer residence north of the region. So the two primary lifting mechanisms which are so conspicuously present in winter are absent in the summer. To be sure, convective lifting continues over the land, but the net result is that the region is much sunnier than during the winter; and the clouds which occur are often the type associated with convective lifting—the fluffy cumulus type, which creates a lively sky with scattered clouds and permits an abundance of sunlight to be received at the surface.

9: The Lake Breezes

FOR NEARLY TWO WEEKS IN THE SPRING OF 1972, the normally active weather systems which affect the Great Lakes were held in abeyance by what the meteorologist calls a *blocking high*. Forecasters explain that blocking in the atmosphere carries the same connotation as it does on the football field. In football, a good blocker attempts to impede the forward progress of the opposition. In the atmosphere, a blocking feature impedes the forward movement of weather systems. Cyclones and anticyclones fail to progress normally from west to east, and stagnation of weather becomes the rule.

Atmospheric blockers take the form of anticyclones or ridges in the upper air circulation. Weather is affected both upstream (to the west) and downstream (to the east). Blocking may cause a favorable weather pattern to persist for days on end. This is an unusual circumstance in the Great Lakes region. More often, the blocking will result in a long period of rainy or drizzly weather or in an extended period of heat and humidity.

During the middle part of May 1972, blocking by an anticyclonic ridge over the eastern United States caused a highly unusual sequence of spring days in the Great Lakes area, each day very much like the next. Each day the temperature rose

127

into the high 70s or low 80s and sunshine was abundant. Each night the temperature fell into the middle 50s—just right for comfortable sleeping. Perfect weather! Unusual for the Great Lakes? Not for a day, or even 2 or 3 consecutive days. But for 12 days in a row? Practically unheard of in the Great Lakes area during spring—not common even during summer in this active weather zone. During the entire period, rain systems lingered to the west in western Wisconsin and Minnesota but failed to penetrate eastward. It was spring in all its glory, drawing people to the fields, parks, lakes, and especially to the Great Lakes.

Along the Lake Michigan shoreline throughout the long spell of repetitious weather, a phenomenon unique to large lakes and water bodies occurred each day. The morning sun rose to greet cloudless skies and practically nonexistent winds. Barely a ripple stirred the glassy surface of the big lake. The pounding of waves on the beaches was replaced by an innocuous lapping of water around buried logs or grotesque driftwood forms. Now and then a fish broke the surface to optimistically sample the morning air.

By 10:00 or 11:00 o'clock each morning a change occurred. The lake, as motionless as a frozen pond during the early morning hours, suddenly broke into a myriad of small ripples. Within minutes, small waves were breaking all along the sandy beach. The air, languid under the increasing warmth of the morning sun, quickly burst into motion. A breeze advanced boldly inland, carrying cooler temperatures and higher humidities. Soon the rustle of leaves and branches and the tinny, hissing sound of miniature sandstorms sweeping along the beach added to the cacaphony of sound accompanying the breeze. An hour later, the cloudless sky was marred by a line of cumuli paralleling the shore about a mile or so inland. The breeze was fresh now, and the sounds of the lake carried into the forest and campground behind the beach. The lake breeze had arrived! It had signaled its arrival and would prevail throughout the day, refreshing the shoreline and tempering the rays of the

Line of cumuli at the Chicago lake front at inner margin of morning lake breeze as it starts inland.

sun. Not a nocturnal creature, its retreat would be gradual and quiet, lacking the brashness of its approach, unnoticed by most people. But it would be gone by evening, and the quiet of the early morning would return.

Along the shores of the Great Lakes, the lake breeze is a noticeable weather phenomenon during 30% to 50% of the days during the summer. The breeze from the lake may extend inland only a few hundred yards or as much as 25 miles (40 kilometers) (Cole and Lyons 1972, p. 442). It results from the same basic water-land temperature differences as other lake-induced weather features. The lake breeze is a stable-season feature, occurring when the lake temperature is colder than the land. Thus the lake breeze has a cooling effect along the immediate shoreline and also increases the humidity. Some very definite weather effects are associated with lake breezes, as well as some definite vegetation associations along the shores.

Local Winds

Lake breezes belong to a group of weather phenomena known as *local winds*. Winds are normally controlled by the positions of the dominant lows and highs; as these features are usually hundreds of miles apart, the winds usually occur at the *synoptic scale*—that is, on a scale of 100–1,000 kilometers. Occasions exist, however, when pressure differences occur within short distances as localized differences in heating and cooling rates cause temperature differences. The winds which result are local winds. They are *mesoscale* features, with dimensions of 100 meters to 100 kilometers. Local winds include land and sea breezes (which occur along the coasts of oceans), lake breezes (which occur along the shores of lakes like the Great Lakes), and a vast array of winds related to topographic differences within small areas, such as mountain winds, valley winds, fall winds, and a variety of upslope and downslope winds.

Local winds have been intensively and extensively studied, and entire books have been written about their occurrence in different parts of the world. The Mediterranean area is particularly noted for the occurrence of local winds, for here a very large water body is surrounded by a great diversity of terrain, with a highly irregular coastline consisting of many mountainous peninsulas and deeply indented bays. Because of the mixture of land versus water and valley versus mountain, local winds with such names as *mistral, scirocco, meltemi,* and *bora* give a colorful meteorological flavor to this area.

In the Great Lakes region, topographic differences are generally minimal, but the breakup of the earth's surface into land and water affords excellent opportunity for a modification of the land and sea breeze cycle to occur. This modified cycle results from differences of seasonal heating and cooling between the land and water, as noted previously, and also from differences in *diurnal* (daily) heating and cooling.

Why Lake Breezes Occur

In the early summer along the shores of the Great Lakes, the average water temperature is considerably cooler than the temperature of the surrounding land. Just as the water undergoes a smaller seasonal change of temperature than the land, so also does the water temperature undergo a much smaller diurnal temperature change. Mixing of the water distributes incoming energy during the day to the deeper layers of water, preventing the attainment of high surface temperatures; while nighttime cooling at the surface decreases the density of the surface water, causing it to sink and mix with deeper, warmer water.

Figure 36 shows a typical diurnal temperature curve for water and land temperatures somewhere around a Great Lake in the early summer. The temperature of the land passes through a cycle of much larger extremes than that of the lake. Notice that, from about 0600 to about 2300 hours, the land is warmer than

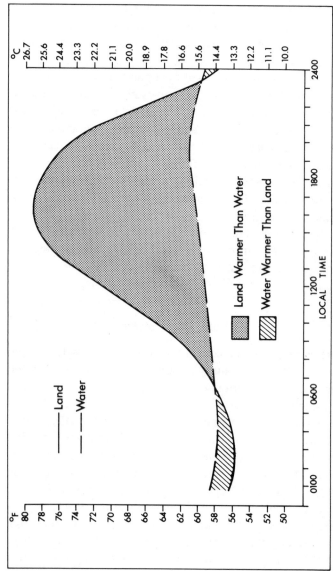

Figure 36   Diurnal Temperatures, Land versus Water, Typical of the Great Lakes in the Early Summer

the lake; and that, from noon to about 1800 hours, it is *considerably* warmer than the lake. From about midnight until around 0600, the lake is warmer than the land.

During the hours when the lake is colder than the land, air passing over the lake from the land is chilled by conduction and radiation so that a shallow *conduction* inversion develops, normally not much more than about 500 feet (150 meters) high. The conduction inversion is typified by cold, dense air near the surface and warmer air aloft. The chilling of the air (and consequent increase in density) causes the surface pressure to rise over the lake, and a mesoscale anticyclone begins to develop, with accompanying subsidence or downward motion.

As the temperature gradient between the land and the lake begins to steepen during the early morning hours, the shallow anticyclone over the lake intensifies, with pressures increasing as much as 2 millibars. On the other hand, the land area is warming quickly; and, since with increasing temperatures the air is less dense, the pressures over the land decrease slightly. Consequently a pressure gradient is set up between the air over the cool water (dense, higher pressure) and the air over the warmed land (less dense, lower pressure). The air should then flow from the lake to the land near the surface, with the return flow aloft. Such a cell-like flow is the lake breeze, which is illustrated in figures 37 and 38.

Dr. Walter Lyons has described the meteorological conditions normally associated with the occurrence of the lake breeze along the shores of the Great Lakes as follows:

A typical lake breeze day has light gradient winds, considerable sunshine, and the afternoon surface air temperature inland rises above the mean temperature of the Lake surface. Generally, a lake breeze will onset several hours after sunrise. Its leading edge, called the lake breeze front, wind shift line, or convergence zone, may penetrate inland anywhere from several blocks to over 40 km. The vertical depth of the inflow layer has been measured by pibal wind

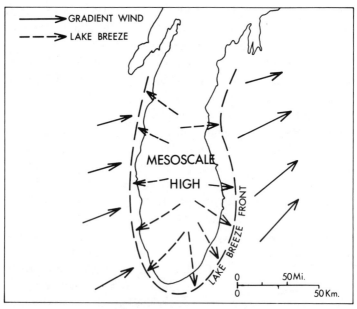

**Figure 37   Lake Breeze, Mesoscale High, and Lake-Breeze
Front over Lake Michigan**

soundings to vary from 100 to 1000 meters, but a value of 400
to 500 meters is most common. The maximum inflow
velocities usually peak at about 5 or 6 meters/second. A
compensating return flow layer aloft tends to have about
twice the depth and half of the speed of the winds in the
inflow layer (Cole and Lyons 1972, p. 442).

If the synoptic scale pressure gradient is such that strong winds
exist over the lake, lake breezes are unlikely to form although
the conduction inversion and lake anticyclone may exist. The
mixing and turbulence associated with the large-scale wind
field will not allow the lake-breeze circulation cell to form. And
if the air passing over the lake is already cool, then little further
cooling will occur and a large temperature differential between
land and lake will not develop during the day. Thus the weather
conditions favoring the occurrence of a lake breeze include light

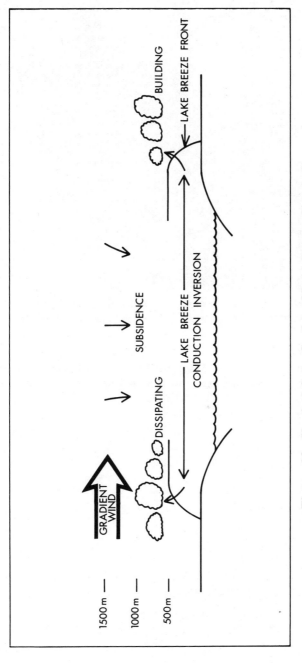

Figure 38 Vertical Cross Section of Lake-Breeze Circulation

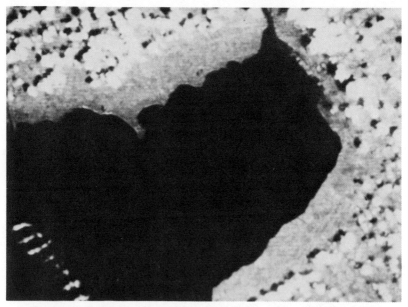

LANDSAT I shows the eastern end of Lake Erie. A line of cumuli marks the lake-breeze front. Note clear zone along the lakeshore.

Figure 39 Diurnal Temperature Curves for Muskegon and Grand Rapids, Michigan, on 26 June 1966, when a Lake Breeze Occurred at Muskegon

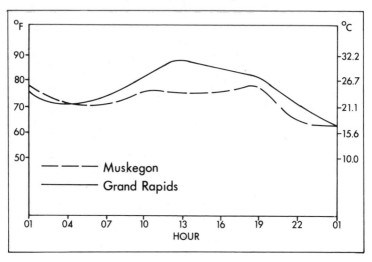

winds, warm temperatures, and plenty of sunshine to warm up the land during the day.

Given its rather frequent occurrence in the Great Lakes area, the lake breeze provides a natural air-conditioning system for lakeshore areas. While a strong lake breeze was occurring in the Chicago area, Lyons noted a temperature of 66°F (18.9°C) at the lakefront; while at Joliet, 30 miles inland, the temperature was 98°F (36.6°C) (Lyons 1966, p. 262). A study of Michigan temperature contrasts between Muskegon on the lake and Grand Rapids 30 miles (48 kilometers) inland on days when lake breezes were in progress showed a distinct flattening of the daily temperature curve at Muskegon as the lake breeze developed (figure 39). The cool temperatures of the lake breeze are accompanied by higher relative humidities and sometimes fog, which may penetrate the immediate shoreline.

Other Weather Effects of the Lake Breeze

Besides altering the temperature and humidity conditions along the shoreline, the lake breeze and its attendant mesoscale anticyclone suppress convective clouds along the immediate shore. Convective clouds form from updrafts of air, or *thermals,* which develop over land areas as a result of intense surface heating of the land. These are the white, fleecy clouds which are so common in summer. They may begin to form around midmorning and become more numerous as the day progresses and the earth's surface continues to warm. During the afternoon, the convective clouds may become thick enough to give some rain showers. The showers are the scattered, local, types which may supply needed moisture to one farmer's field but supply no rain whatsoever to the farm down the line.

Over inland portions of the Great Lakes basin, the buildup of cumuli on summer days is a very common sequence. Along the immediate shores of a lake, however, the sky may be free of cumuli, and the same may be true over the lake itself. Evidently some mechanism is operating to suppress the growth of this cloud type near and over the lake. The mechanism responsible is the intense cooling of air below the conduction inversion which restricts vertical motion and the subsidence associated with the lake anticyclone. Also, when the lake-breeze front penetrates inland, the cool air behind the front effectively counters the convective uplift which would otherwise occur during the day. The immediate shore area behind the lake-breeze front is usually free of clouds. Photographs taken from satellites clearly show this cloud suppression on days when lake breezes are strongly developed (Lyons 1966). On many days, one can look landward while standing on one of the Great Lakes shores and see the sharp demarkation of the lake-breeze front in the form of a wall of cumuli, which appears to remain almost stationary throughout the day. That's when resort owners on the shores of the Lakes seem almost in conspiracy with the lake breezes to make the shoreline climates cooler and sunnier on hot summer days. Occasionally, however, the villain fog enters the picture when the lake breeze overplays its hand.

The lake-breeze front in itself may become a weather-active zone. It is a zone of convergence, marked by an upward motion in the atmosphere. Occasionally a line of rather persistent showers will develop along the lake-breeze front. The showers, although mostly light, will tend to remain positioned at the inner margin of the lake breeze. The author has noted occasions when, about 20 miles (32 kilometers) inland from the eastern shore of Lake Michigan, intermittent showers descended throughout much of the afternoon from a deck of clouds which lingered directly overhead and covered the entire eastern half of the sky. At the same time throughout the afternoon, the western quadrants remained completely free of clouds!

What effects do the cold waters of the Great Lakes and the lake-breeze fronts have on squall lines or severe weather systems? The answer is not quite so clear. Some thunderstorm lines seem to weaken and nearly dissipate upon moving over the

Cloud line as a result of a double lake-breeze convergence over the eastern portion of Michigan's upper peninsula.

cold lake. In other cases, the convergence along the lake-breeze front seems to intensify existing weather systems, and severe storms may result. Much depends upon the individual weather situation and the alignment and intensity of the squall line, as well as the existence or absence of a lake breeze. In the Detroit metropolitan area, some evidence exists that the lake-breeze front which pushes westward from Lake St. Clair (yes, even a lake that small is capable of generating lake breezes) may be at least partially responsible for the occurrence of some excessively heavy rains over the northeastern suburbs of the city.

One other interesting feature associated with lake breezes should be noted. A glance at the map will show that the eastern end of Michigan's upper peninsula is rather narrow. To the north is Lake Superior; and on the south, not much more than 40 miles distant, is Lake Michigan. Both of these lakes are capable of generating well-defined lake breezes. What happens if lake breezes from both lakes should occur at the same time? For years, people vacationing during the summer in the Straits of Mackinac area have noted the persistent line of clouds which seemed to hover far to the north over the interior of the upper

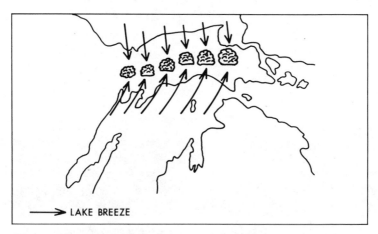

**Figure 40   Double Lake-Breeze Convergence over the Eastern
Upper Peninsula of Michigan**

peninsula. At the same time, all other areas would be cloud-
free. The line of clouds is caused by the double lake-breeze
convergence as lake-breeze fronts from Lake Superior and
Lake Michigan push inland and meet along a line somewhere in
the middle of the peninsula (figure 40). Showers may be
generated along this line, and the amount of sunshine during the
summer is reduced.

10: Lake-Effect Snow

SOME CURIOUS AND UNIQUE WEATHER OCCURS ALONG the lee shores of the Great Lakes during the winter. Brief, blinding snow showers are punctuated intermittently by periods of brilliant sunshine. The snow descends from angry, dark, moisture-laden clouds which float onshore from the Lakes. The snow showers extend only 20 or 30 miles (32–42 kilometers) inland, however. Areas close to the Lakes may experience heavy snowfalls, but inland areas seem to escape the brunt of these storms. Sharp dividing lines often mark the edges of the storm systems.

The author well remembers an automobile trip from Kalamazoo, in southwestern Michigan, to Chicago on a wintry day in 1958. Light snow had occurred in Kalamazoo throughout the night, although the total accumulation was not large. Approaching the Lake Michigan shoreline from the east, the sky darkened and much heavier snow squalls were encountered. The highways became buried, and it was obvious that a much deeper blanket of the white stuff had fallen during the night along the lakeshore than inland.

Near Michigan City, Indiana, the scene was chaotic. Roads were nearly impassable, cars were buried, and mailboxes were completely covered. Heavy snow squalls were sweeping in off Lake Michigan from the north, and little hope appeared of ever reaching Chicago on that day.

Suddenly, just to the west of Michigan City, the scenario changed. The sun burst forth, and within another 1½ miles, the ground was bare! There was little evidence of the ferocious snowburst which had buried the southeastern tip of Lake Michigan under as much as 50 inches (127 centimeters). Looking back to the east, however, the edge of the storm appeared as a line of towering clouds resplendent in the morning sunshine. The bases of the clouds, from which the snow showers were descending, were ominously dark, while the tops resembled the shining turrets of summer thunderstorms. The author returned from Chicago later in the day and again encountered the same line-up of clouds near Michigan City. It had not changed position, and it warned of the deep snows and bad driving conditions which lay ahead. Within 1½ miles, the bare landscape was again buried!

Such seemingly "freak" snowstorms are not at all unusual to the Great Lakes. For example, on 10 December 1937, a snowstorm buried Buffalo's north side under 2 to 4 feet (61–122 centimeters) of snow, with drifts to depths of 8 feet (244 centimeters). Hundreds of stranded automobiles were completely covered by the drifts, and many persons on their way home from work were trapped by the storm and forced to spend the night in unfamiliar surroundings. Little snow, however, occurred on the south side of the city. B. L. Wiggin (1950), for many years the meteorologist-in-charge of the U. S. Weather Bureau station at Buffalo, has described the sudden snowstorms which whirl in from Lake Erie. As much as 50 inches (127 centimeters) of snow may fall in a narrowly concentrated zone from these highly localized storms.

Oswego, New York, near the lee end of Lake Ontario, has had some legendary snowfalls. The famous snowburst of 7–11 December 1958 resulted in the accumulation of 66.7 inches

Winter of 1976–77, Adams, New York, after a series of lake-effect storms.

(170 centimeters) within a 200-square-mile (517 square-kilometer) area. And the 5-day total for 27–31 January 1966 at Oswego was a whopping 101 inches (256 centimeters)! There could have been no more appropriate site for the Twenty-ninth Annual Eastern Snow Conference than Oswego; when the conferees met on 3–4 February 1972, they were greeted by as much as 50 inches (127 centimeters) of snow and stranded in Oswego until the morning of February 6.

More recently, in February of 1976, Barnes Corner, New York, at the eastern end of Lake Ontario, experienced a snowburst of 54 inches (137 centimeters)in less than 24 hours. During the same month, a snowburst south of Buffalo dropped 42 inches (107 centimeters), and the winter of 1976–77 produced prodigious amounts of snow in the Buffalo and Watertown areas of New York. Muskegon, Michigan; South Bend, Indiana; Erie, Pennsylvania; and Syracuse, New York; have their snowbursts too. And even the Windy City, Chicago, situated on what is normally the upwind shore of Lake Michigan, occasionally experiences them.

Such highly localized snows, occurring along the immediate downwind shores of the Lakes and extending inland for 25 or 30 miles, are called *lake-effect snows*. A large percentage of the annual totals of snowfall in certain Great Lakes areas comes from these localized lake-effect systems, and the heavy snows during the winter of 1976–77 were mostly lake-effect in origin. Lake-effect snowfall may be of serious consequence to the inhabitants of the Great Lakes area, accounting for countless lost days of work and school, making driving hazardous, wearing out snow shovels and shovelers—in general, making winters even more wintry than they already are. But to the ski resort operator or snowmobile owner, lake-effect snows are additional Christmas gifts delivered by nature. And that feeling of excitement which children and even grownups share on waking to the first snow of winter is often the result of a direct contribution by the Great Lakes.

Why Lake-Effect Snowstorms Are Unique

Obviously, all snowfall in the Great Lakes area does not result from lake-effect snow. Passing cyclones contribute their share, and snowfall amounts from these systems are much less affected by the presence of the Great Lakes. But the lake-effect storms develop when the surface weather map indicates no apparent cause for a snowstorm to occur. They are unique in the absence of any cyclone or front in the area; in fact, the normally fair-weather system, the anticyclone, may be the dominant pressure feature.

Lake-effect storms are mesoscale features, generating on a smaller scale than cyclones and anticyclones. Data that describe processes occurring at the mesoscale are difficult to gather with the present spacing of the National Weather Service network. Even snowfall distributional patterns may escape detection by snow-measuring stations. These storms occur on such a small scale that heavy snows between stations may go unrecognized except by the residents of these areas. So special, more dense instrumental networks must be employed to inform us about these storms.

Lake-effect snowstorms are also unique in their lack of occurrence in most other areas of the world. Thus Great Lakes residents have a weather feature they can truly call their own. Just the right combination of geographic factors must be present for these storms to occur: A large water body must be located upwind of the area to supply the necessary modifications; the area must be located far enough from the equator so that winter temperatures are sufficiently cold for precipitation to occur as snow—but not so cold that the body of water which modifies the air will freeze over completely; the body of water must be large enough to warm the air, but not large enough to warm it above the freezing point; and a land mass of continental dimensions must lie upwind to supply the cold air. Thus the requirements necessary for these types of storms to form on a significant scale are quite restrictive—and other than the Great Lakes area of

Cloud line with snow squalls over Lake Huron, photographed from a NCAR research aircraft.

North America, they occur only along the east shore of Hudson Bay in Canada and along the west coast of the Japanese islands of Honshu and Hokkaido.

The Causes of Lake-Effect Snowstorms

Over half a century ago, meteorologist C. L. Mitchell queried several vessel masters of Pere Marquette ferries plying the waters between Ludington, Michigan, and Milwaukee and Manitowoc, Wisconsin, regarding the behavior of weather over Lake Michigan during the winter. These individuals crossed and recrossed the lake under all sorts of conditions and had firsthand observational knowledge of conditions miles from shore. One of the skippers, a Master Bahle, replied:

"With the wind west and weather clear we may have vapor or steam, as we call it, part or all the way across the Lake. All depends on the difference of temperature of the water and air.

During the early part of winter, say in December, when water is not the coldest, the weather will moderate as we reach the east shore, and this will cause the steam [fog] to rise off the water entirely in clouds and then snow may fall. I have seen this anywhere from the middle of the lake to the east shore. Later in the winter when the water becomes real cold and the air temperature say about 15°F below zero on the west shore, the steam [fog] may reach the east shore and snow there also, the snow not extending out in the lake more than 2 or 3 miles. In other words, it does not snow when the steam [fog] makes. It starts to snow where the steam [fog] stops making and starts to rise in clouds entirely away from the water." (Mitchell 1921)

Other vessel masters wrote that fog rising on the western side of the lake meant snow on the eastern side and that these flurries extended back 10 to 20 miles from the eastern shore, as a rule.

The replies of the vessel masters imply that somehow the winter fog (steam fog) rising from the Great Lakes is a harbinger of snow on the downwind shore. If their observations were correct, one may assume that the processes which form the fog and the snow are linked. This is actually the case, and in order to understand what causes the two phenomena, we must once again consider the relative temperature of the Lakes and the land during the winter season.

As previously noted, the unstable season on the Great Lakes lasts from middle or late August until about 1 March. The maximum mean temperature differences between land and water occur from about middle or late November until middle or late January. During this time, temperatures of the southern Lakes may average over 15°F (8.3°C) warmer than the land, while the northern lake temperatures may average 30°F (16.6°C) warmer than the land.

But this doesn't tell the whole story. Occasions exist when very cold air from the Arctic penetrates well southward into the

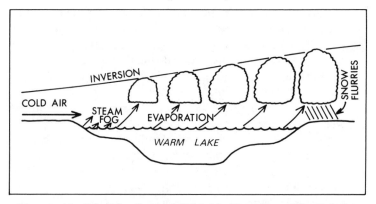

Figure 41   Modifications of Cold Air Crossing a Warm Lake

United States, sweeping across the Great Lakes on its path to the Atlantic and to the Gulf coast. Under these conditions, the contrast in temperature can become considerably larger, as much as 40°F (22°C) in the south and 50°F (28°C) in the north. With such large temperature differences, the Lakes interact with the air in decisive fashion.

As the cold air streams across the warm Lakes, it is warmed and becomes more humid. As the air warms, it becomes less dense and tends to rise, cooling adiabatically. Whenever moist air rises, as previously noted, clouds may form and precipitation may result. Hence the observations of the ferry masters concerning "steam" (fog) and snow. The fog results from the intense evaporation or transfer of moisture from the warm water to much colder air when the cold air initially makes contact with the warm water. After the air passes for some distance over the lake, convection and turbulent exchange have transported the acquired moisture aloft to form clouds, and snowfall may occur (figure 41). Hence the statement by Master Bahle that "it does not snow when the steam [fog] makes. It starts to snow where the steam [fog] stops making. . . ."

As the warmed air reaches the shoreline, additional lifting may be induced as the air begins to "pile up" because of the

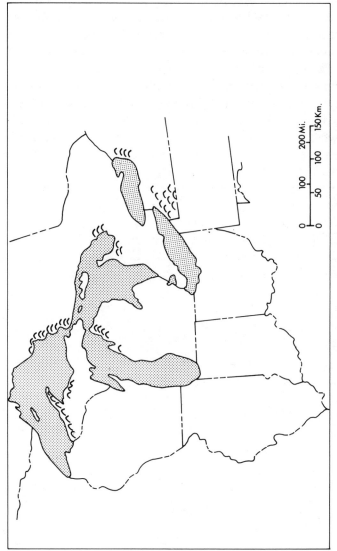

Figure 42   Hilly Areas or High Ground to the Lee of the Great Lakes

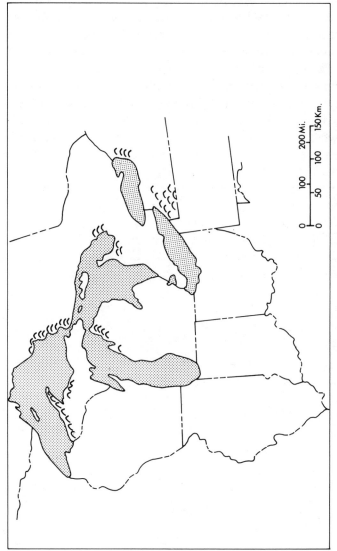

increased friction over the land area. And with the presence of hills and high territory on the downwind shores of the Lakes, an additional impetus for cloud formation and snowfall is provided by orographic lifting. Rather high hills occur along the east shore of Lake Superior, on the Keweenaw peninsula of Upper Michigan, in northwestern Lower Michigan, at the eastern end of Lake Ontario, and inland from the southeast shore of Lake Erie (figure 42). While the Lakes will cause clouds and snowfall to occur without the presence of hills or high ground, rough terrain may amplify the snowfall amounts.

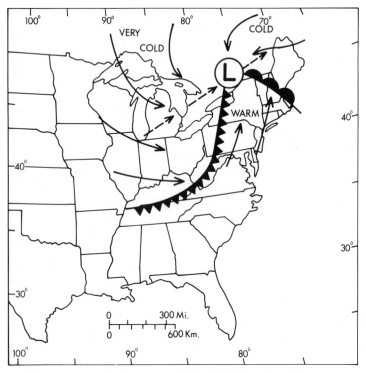

Figure 43   An Outbreak of Very Cold Air May Follow the Passage of a Cyclone through the Great Lakes

Usually the incursions of Arctic air necessary for large amounts of lake-effect snow occur after a strong cyclone has passed through or near the Great Lakes area. When the cyclone moves through, it opens the way for cold air to rush southward, usually in the form of a high-pressure area behind a cold front. Thus the snowfall usually occurs in conjunction with a rising barometer. Although the cyclone may have progressed far east of the region and may, in fact, be situated off the East Coast or over the Gulf of St. Lawrence, it continues to exert its control over the Great Lakes by causing the outpouring of very cold air (figure 43).

As the cyclone varies in its location, it "wags the tail" of the winds, whose direction over the Lakes is controlled by the positions of the nearby high-pressure and low-pressure areas (figure 44). Above the surface and not evident on the surface chart, a favorable lifting environment for the air is provided by the winds.

Do humans play a role in initiating lake-effect snowstorms?

The Great Lakes modified the blasts of Arctic air from Canada long before people cut down the forests, plowed the land and built large cities. Thus there is little doubt that lake-effect snowstorms have been occurring for centuries. But during recent decades, portions of the Great Lakes have become heavily populated and industrialized, and major cities have appeared.

Meteorologists know that cities, even small ones, alter weather and climate in many ways. We will discuss the role of cities in more detail later, but it is presently interesting to note that cities are often warmer than their surrounding areas; and that the atmosphere acts as a huge "garbage can," removing the gaseous and particulate waste products which enter the air because of man's activities. Thus the air passing over large urban areas becomes warmed and also acquires an increased

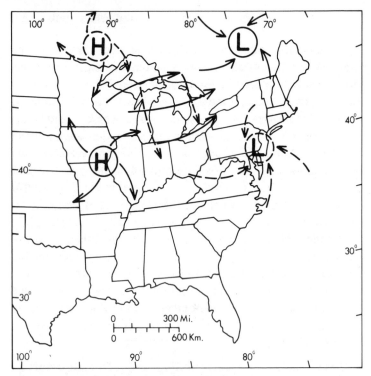

**Figure 44 The Positions of the Dominant High-Pressure and
Low-Pressure Areas Determine the Direction and Fetch
of Winds across the Great Lakes**

load of pollutants. The warmth from the cities may add to the
heat acquired by the Lakes and may occasionally provide a
stimulus for development of lake-effect snowstorms.

And city industries may also inject what are called ice-
forming nuclei into the atmosphere. These are special types of
particulates which may play a role in the formation of ice
crystals and snowflakes. Normally the atmosphere does not
contain an overabundance of ice-forming nuclei, so the quantity
introduced by man's activities may become significant in
certain areas. Steel mills are known to introduce ice-forming
nuclei into the atmosphere, and the southern Great Lakes area is
one of the world's leading centers for the manufacture of iron

and steel! In addition, lead from automobile exhausts combines with the natural iodine in the air to form lead-iodine compounds, and these may also become active in initiating the growth of ice crystals.

So we can see that the large urban areas along the shores of the Great Lakes may, at times, play a role in creating or intensifying downwind lake-effect snowstorms through the additional heat and ice-forming nuclei which they supply.

Where and When Will Lake-Effect Snowstorms Occur?

Essentially, the greater the temperature difference between the lake and the air, the larger the potential for lake-effect snow. And the proper weather pattern must exist so that a favorable situation for lifting occurs within the lower portions of the atmosphere.

Probably of greatest importance, however, in determining which areas will be affected is the *fetch,* or the length of wind travel over the open water surface. The longer the fetch, the greater the amount of heat and moisture which may be acquired from the lake; and consequently, the greater the potential for lake-effect snow.

In order to understand how fetch may be important, let's reexamine the map of the Great Lakes (figure 1). Notice that the shape and orientation of each lake is somewhat different. Lakes Michigan, Erie, and Ontario are elongated in shape, resembling cigars or dirigibles. Each has what might be identified as a long axis and a short axis. Note, however, that the long axis of Lake Michigan is oriented in a north-south direction, while the long axes of both Lakes Erie and Ontario are aligned in a west to east direction.

Thus if winds are prevailingly from the north over the Great Lakes, they sweep down the long axis of Lake Michigan—a fetch of nearly 300 miles (483 kilometers)—but over Lakes Erie and Ontario, the winds cross the short axis, with a fetch of

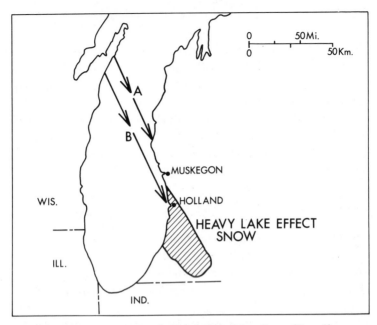

Figure 45 Variation in Fetch Resulting from Shoreline Configuration

only about 60 miles (97 kilometers). The air sweeping over Lake Michigan has a very large fetch and can acquire large amounts of heat and moisture before crossing the downwind shoreline. Winds crossing Lakes Erie and Ontario have less of an opportunity to do so. However, winds from the west or southwest cross only about 60 miles of Lake Michigan, while they may parallel the long axes of Lake Erie and Lake Ontario.

Lake Huron and Lake Superior, while not quite so elongated in shape as the other Lakes, nevertheless have some narrow parts and wide parts, and fetch cannot be neglected as a factor.

Along the shorelines of the Great Lakes, the alignment and configuration of the shore can also be a major control in restricting lake-effect snows, or in concentrating them in certain areas. For example, note the configuration of the shoreline along the eastern side of Lake Michigan from near Holland to about 50 miles north of Muskegon, Michigan (figure 45). The shoreline trends a little west of north along this stretch, but the alignment changes both to the north and south. Given this particular shoreline orientation, winds blowing across the lake from the north-northwest may have only a 60-mile fetch at point *A* but a 150-mile fetch at point *B*. Under such conditions, lake-effect snowfall may be rather light along, and inland from, the shore at point *A* but may be very heavy at point *B*. Thus the snowfall may be heavily concentrated in the counties of southwestern Michigan, and a sharp breakoff may be encountered as one travels from the area affected by a lengthy fetch to an area affected by a much smaller fetch.

Fetch may also determine whether the winds pass over large industrial areas, receiving additional heat and ice-forming nuclei. Figure 46 shows a situation in which the winds were blowing across Lake Michigan from the southwest. The air was much colder than the water, even though the winds blew from a southerly quadrant. This was air of Arctic origin which had recurved in its travels from northern Canada until it was actually approaching the Great Lakes area from the southwest (with special types of weather situations, this is entirely possible).

As the air currents overpassed the warmer lake, heat and moisture were acquired, clouds formed, and snow squalls began to break out on the downwind shore. The snowfall was heaviest in the area directly downwind from the Gary–South Chicago industrial area—an area noted for the manufacture of steel! Although it is impossible to say whether or not the industrial area significantly affected the distribution of snowfall, the circumstantial evidence from situations like this one certainly suggest that this may be true.

Snow bands

Fetch may also determine whether the snowstorms will be in the form of *bands* of clouds. These often form when the

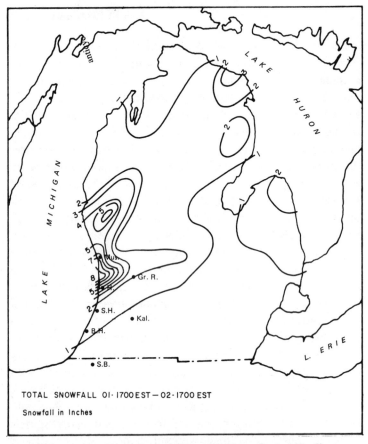

**Figure 46 Total Snowfall for 24-Hour Period Ending
5:00 P.M. EST, 2 January 1970.**

winds are aligned parallel to the long axis of the lake. The resulting linear banding of clouds can often be seen on satellite photos (p. 101).

The banding is also evident when the precipitation areas are viewed by radar. Radar has the capability of "seeing through" clouds, but areas where precipitation is occurring show up as reflective "bright spots" on the radar scope. This is because the wavelengths of the signal emitted from the radar are calibrated to bounce (echo) off large droplets or ice crystals, thus being

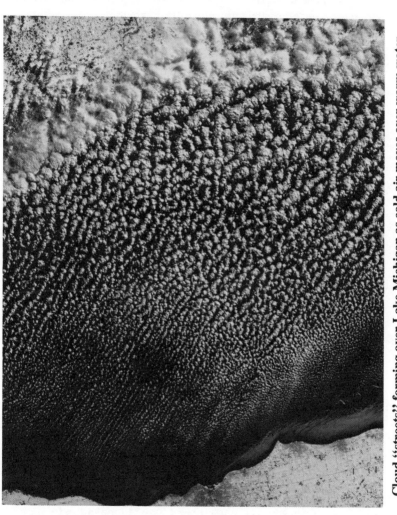

Cloud "streets" forming over Lake Michigan as cold air moves over warm water. A LANDSAT I photo.

reflected back to the receiver. Small water droplets or crystals will not have this effect. While the ordinary cloud contains small droplets or crystals, the cloud from which precipitation is occurring or likely to occur will contain large droplets or crystals. Thus the echo, or reflected image on the radar scope gives a fairly good clue as to where the precipitation is. The radar can be very useful in "picking out" lake-effect storms, which may occur on too small a scale to be detected by the regular observing network. The heaviest precipitation seems to occur on the terrain immediately below the origin of the echo.

The bands of cloud and snow identified by radar or satellite may range in width from about 2 to 20 miles (about 3–32 kilometers); and in length from 50 to 100 miles (80–160 kilometers). The heaviest snows fall from the narrow bands where they intersect the shoreline and for about 20 or 30 miles (about 30–40 kilometers) inland. Outside the bands, no snow at all may be observed. The bands have the capacity to maintain a steady state for hours, dumping all the snow into a narrowly defined zone beneath the band. Such was the case along the southeast shore of Lake Michigan on the night of 20–21 January 1970 (photo, p. 159). This band was oriented parallel to the flow of air over Lake Michigan and intersected the shoreline in the area of Benton Harbor–St. Joseph, Michigan. As much as 8 inches (20 centimeters) of snow fell within a period of 6 hours. Then the conditions changed so that uplift of the air in the lower layers was no longer supported. The snowfall quickly stopped and the band disappeared from the radar scope within half an hour.

Such banded snowstorms are responsible for most of the heavy lake-effect snows along the lee-shore areas. The Michigan City case and the snowburst in Buffalo, New York, described earlier were both of this nature. Banding may occur with north or northwest winds over Lake Michigan, and with west or southwest winds over Lakes Erie and Ontario. Over Lakes Superior and Huron, banding may occur with a greater variety of wind directions.

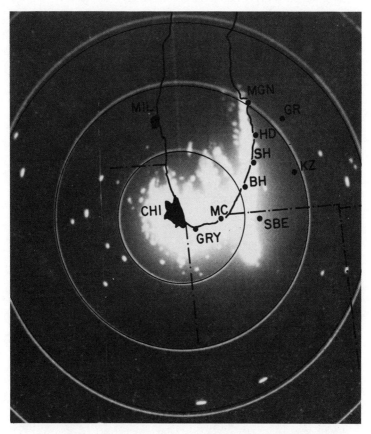

A strong lake-effect snow band parallels the Lake Michigan shore, extending onshore near Benton Harbor, Michigan. Chicago Radar 1002 EST, 20 January 1970.

But lake-effect snow can also occur when banding is not so evident, as when the flow is cross-lake or parallel to the minor axis. Satellite pictures then tend to show a cloud deck consisting of parallel cloud streets or lines which may develop only 10 or 15 miles (16–24 kilometers) from the upwind shore. Flurries of snow may occur all along the downwind shore and may be heaviest where hilly areas exist, thus contributing a little additional uplift. The satellite photo on p. 101 shows where such a cloud deck is evident over Lake Michigan and its eastern shore. The winds were mostly westerly when the picture was taken. With a more limited fetch because of cross-lake flow, organized individual snow bands did not develop—but with cold air crossing the lake all along its 300 miles (483 kilometers) of length, a more continuous cloud deck along the eastern shore resulted.

The Climatology of Lake-Effect Snowfall

While lake-effect snowfall may be viewed in different ways by different people, one saving feature about it pleases practically everyone. The snow is usually very light and fluffy with a low water content. Thus it is easy to shovel off driveways and porches and provides the type of surface that skiers like.

In order to explain what is meant by low water content, we must briefly consider the techniques employed in the measurement of snow. Probably no weather element causes more measurement problems than snow. The difficulties arise from the fact that the wind blows the snow into drifts and gullies, tending to remove it from the flat, open surfaces. It is also rather difficult to catch snow in gauges, as is done with rain. Therefore in measuring the depth of freshly fallen snow, the observer is best advised to take a number of measurements with a ruler or measuring stick over a flat surface where drifting is minimal, then to average the measurements. Occasionally a *snow board* is used—that is, an artificial flat surface on which the snow can

Some of the heaviest annual snowfalls east of the Rockies are received on Michigan's Keweenaw Peninsula.

accumulate, be measured, and then swept clear to await the next snowfall.

In addition to the depth of freshly fallen snow, the observer should determine the water content. This can be done by inserting a cylinder or tin can into the snow so that a cross section of the snow accumulation is obtained. Then the snow in the container is melted and the depth of water measured. The ratio of the measured water to the measured depth of snow is called the *water ratio,* and the measured amount of water is the *liquid equivalent.**

As snow usually consists of clumped-together ice crystals, it contains many air spaces. This is one reason why snow is a good insulator. The very light, fluffy snowfalls contain much air space and may have water ratios of twenty-five or thirty to

*In Canada, the liquid equivalent is estimated by using a ratio of ten to one at many stations. Larger stations (synoptic stations) catch the snow in shielded gauges, then melt and measure it for water content. In the United States at both climatological and synoptic stations, the liquid equivalent is determined by melting snow caught in a gauge.

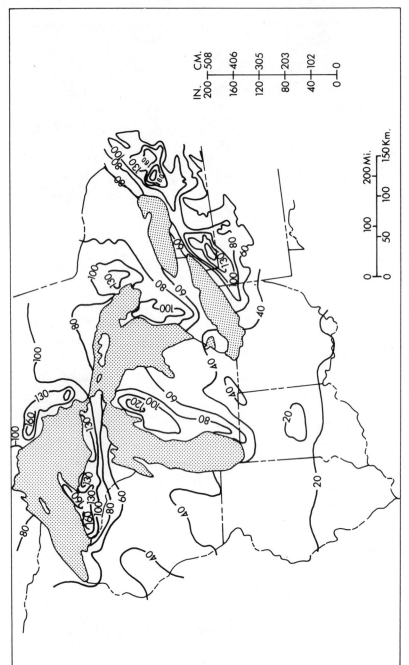

Figure 47 Mean Annual Snowfall in the Great Lakes Area (in inches)

one. This means that if 30 inches (76.2 centimeters) of freshly fallen snow is melted, only 1 inch (25.4 millimeters) of water would remain. On the other hand, some of the cyclonic storms passing through the Great Lakes deposit snowfalls with a very high water content—perhaps four or three to one. These are the wet, sticky snows which are so heavy and difficult to shovel from sidewalks and driveways. The ratio means that we need melt down only 3 or 4 inches of this stuff to get 1 inch of water!

Being largely air, the lake-effect snowpack tends to compress rather rapidly. The freshly fallen lake-effect snow may be 6 inches (15.2 centimeters) in depth if measured very quickly after the snowfall has occurred. However, after a day or so, the snowpack has compressed due to its own weight, and the depth may measure only 3 or 4 inches.

Annual snowfall totals around the Great Lakes

With lake-effect snowfall occurring every year in the lee-shore areas, the annual amounts of snow in the Great Lakes area vary greatly from place to place. As you might guess, there is a general tendency for snowfall to increase toward the north, where winters are colder and most precipitation is in the form of snow. But another pattern is also discernible. The amounts of snow along the downwind shores of the Lakes are considerably greater than on the upwind shores or in interior areas (figure 47).

The downwind shores, where lake-effect snows are frequent, are known as *snowbelts* with annual totals considerably higher than elsewhere. Each of the Great Lakes has its individual snowbelt or snowbelts (figure 48). The presence of the snowbelts means that some of the southern portions of the Great Lakes may have significantly greater snowfall amounts than some of the northern portions. For example, southwestern Michigan counties located near Lake Michigan have higher annual snowfall totals than some counties in the upper peninsula that are located a considerable distance from Lake Superior. And the lee shore of Lake Erie, in southwestern New York and northwestern Pennsylvania, may have heavier annual snow totals than the city of Thunder Bay, Ontario, along the north shore of Lake Superior. Of course, the snow at Thunder Bay is likely to remain on the ground continuously throughout the winter, while the New York–Pennsylvania area is subjected to frequent spells of thawing weather.

Wherever there are hills or higher elevations, the snowbelt totals may be even greater. Snowfalls on the Keweenaw Peninsula of Upper Michigan have been described by A. H. Eichmeier (1951), former State Climatologist for Michigan, as "Paul Bunyan style." The peninsula juts into Lake Superior so that winds from the west, north, or east have a lengthy fetch over the lake before striking the land. And the peninsula consists of a spine of hills and high ground reaching almost 1,000 feet above the level of the lake. Thus in the winter, snowfall is extremely heavy on the peninsula, averaging over 200 inches (500 centimeters). During one year, 1950, snow fell for 51 consecutive days.

The Keweenaw Peninsula's claim for being the snowiest nonmountainous location in the eastern United States is challenged, however, by the towns on the Tug Hill Plateau at the eastern end of Lake Ontario. A sign outside the town of Boonville, New York, proclaims it as the "snow capital of the eastern United States." To be sure, the snow lies long and deep at Boonville. But Bennett Bridge, about 30 miles east of Oswego, New York, may be even snowier: On 17 January 1959, 51 inches (130 centimeters) of snow in 16 hours buried Bennett Bridge. And during the 1946–47 season, 352 inches (894 centimeters) of snow were recorded there. But the winter of 1976–77 shattered all records in the Lake Ontario area. Many stations recorded new seasonal records and Hooker, New York (southeast of Watertown), recorded 466.9 inches (1,185.9 centimeters) of snow—the greatest seasonal total ever listed east of the Rockies!

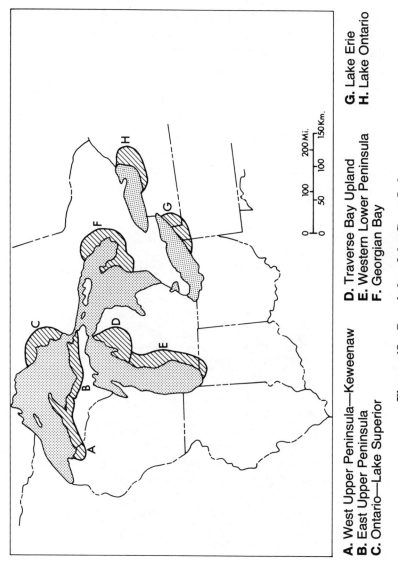

A. West Upper Peninsula—Keweenaw D. Traverse Bay Upland   G. Lake Erie
B. East Upper Peninsula   E. Western Lower Peninsula H. Lake Ontario
C. Ontario—Lake Superior   F. Georgian Bay

Figure 48   Snowbelts of the Great Lakes

The eastern Lake Ontario region is not nearly as far north as the Keweenaw Peninsula, and to understand why it receives so much snow, the topography must be examined. The Tug Hill Plateau slopes upward from the eastern shore of Lake Ontario until it reaches elevations slightly in excess of 2,000 feet (608 meters). When winds are from the west and the air is very cold, as is often the case in that part of New York State, the air travels a long distance over Lake Ontario, gaining heat and acquiring moisture. Lake-effect snow bands may develop. At the eastern end of Lake Ontario, the winds are forced to ascend the Tug Hill Plateau. This forced rise of air is an added factor to the uplift which occurs as the air flows over the warm lake. Thus

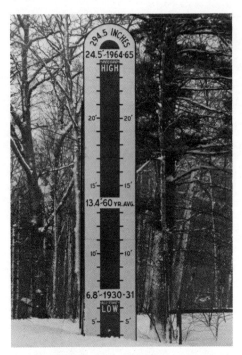

**Snowfall pole on the Keweenaw
Peninsula of Michigan.**

Herman, Michigan. Its elevation and proximity to Lake Superior make Herman subject to record snowfalls.

the Tug Hill Plateau receives a double-barreled blast of lake-effect snow—and substantiates its claim of being the snowiest place east of the Rockies.

In all the snowbelts of the Great Lakes where cyclonic storms and lake-effect snows combine to whiten the landscapes each winter, extraordinary seasonal snowfall amounts have been recorded. In addition to the remarkable accumulations in the Tug Hill area, Old Forge, New York, just within the Great Lakes basin, recorded 408 inches (1,037 centimeters) in 1976–77. Steep Hill Falls, Ontario, at the eastern end of Lake Superior, recorded 301 inches (764.5 centimeters) during the 1939–40 season. And in Michigan, Herman had a total of 308.4 inches (783.3 centimeters) during the 1975–76 season, achieving a new record for that state—until, during the very next winter, Tahquamenon Falls received 332.8 inches (845.3 centimeters). Such amounts make the Great Lakes extraordinary snow factories!

Early winter is the big season for lake-effect snow

As the lake surface temperatures become colder and the lake becomes more extensively ice-covered in late winter, the frequency and intensity of lake-effect snowfalls taper off. Thus November, December, and January are the big snow months, while amounts diminish in February and March. East of Lake Erie, a marked drop-off occurs after early January as the lake becomes ice-covered. As the remainder of the Lakes remain mostly ice-free, the drop-off is not so drastic but is still noticeable in other snowbelts. In Michigan, surrounded by three of the Great Lakes, December snowfall is considerably greater than the February snowfall (figure 49), since February is subjected to less frequent occurrences of lake-effect snow. Lake Ontario, a deep lake, remains relatively ice-free throughout much of the winter and is capable of generating severe lake-effect snows even during the late winter months.

Another feature concerning lake-effect snow was noted by climatologist Charles F. Brooks back in 1914. He wrote:

> Early in winter the snowfall of the immediate shores is generally less than that at a short distance inland . . . late in winter the snowfall of the shore equals that of the higher land. (Brooks 1914)

The retreat of the zone of maximum lake-effect snowfall from the interior to the lakeshore as winter progresses has been examined by Norton D. Strommen, former State Climatologist for Michigan and now Director of the Environmental Data Service's Center for Climatic and Environmental Assessment. He found not only a movement from inland to lakeshore in western Michigan from November to January but also a return movement inland by March. This cycle was most closely related to seasonally changing wind speeds at about the 5,000-foot level. As the winds decreased in velocity during midwinter, the zone of maximum lake-effect snow moved to within 10 miles of the Lake Michigan shore. With higher velocities during early and late winter, the snow was deposited farther inland (Strommen 1975).

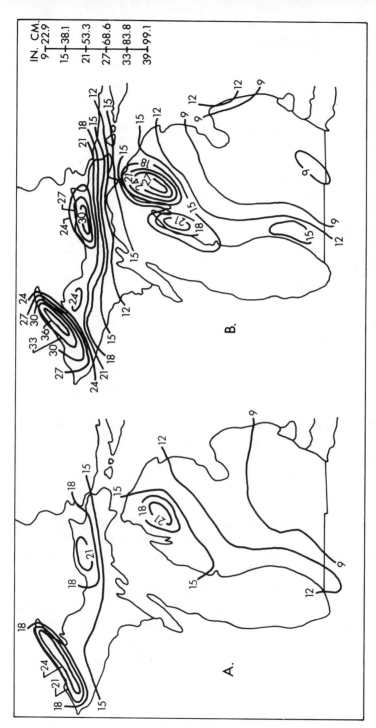

Figure 49 Mean February (A) and December (B) Snowfall for Michigan (in inches)

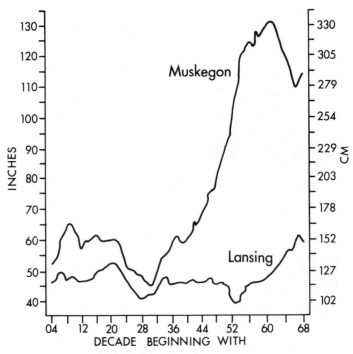

Figure 50   Ten-Year Moving Averages of Seasonal Snowfall, Muskegon and Lansing, Michigan

Are lake-effect snowstorms becoming more common?

As potent as lake-effect snowbursts can be and have been, evidence exists that their frequency and intensity have increased during recent decades. Comparisons have been made between snowfall totals during this century at stations where lake-effect snowfall is not a factor and at nearby stations where it does occur. While the non-lake-effect stations frequently have shown totals that varied little in the course of the century, the lake-effect stations, particularly in southwestern Michigan and to the lee of Lakes Erie and Ontario, have shown sharply increased totals—especially during the 1960s.

In southwestern Michigan, snow amounts have doubled and, in some cases, almost tripled since the earlier decades of the century. For example, compare the amounts received by Muskegon along the shore of Lake Michigan and Lansing in the interior of the state (figure 50). Muskegon's totals have nearly tripled in this century, while Lansing's have remained nearly constant. Some indications exist that the trend toward increased lake-effect snowfall may have leveled off or reversed, but it is too early to say for sure—and the winter of 1976–77 brought lake-effect snow with renewed intensity.

What are the possible causes for these large increases? It is evident that winters have been growing colder throughout much of the Great Lakes area, at least until the early 1970s. Accompanying this cooling trend were more frequent invasions of cold air from the Canadian Arctic. Thus the situations during which lake-effect snows occur were becoming more common. And supplementing the effect of this cooling trend were increases in urbanization, industrial activity, and the numbers of automobiles in the area. Probably all of these factors, as well as others not yet recognized, played a role in accounting for the increase of lake-effect snowfall. At any rate, within many of the Great Lakes snowbelts, the old adage that snowfall was much heavier in grandma's day is quite untrue.

Economic impact of lake-effect snowfall

Robert Muller, a geographer, has surveyed some of the economic consequences of heavy lake-effect snowfalls, particularly as they occur within New York State. He noted that heavy snow could be either an economic gain or loss for local political units, as subsidies of state funds filtered downward to local levels when heavy snows occurred. During a winter with little snow, however, a large investment in equipment and manpower is inactive and adequate funding for salaries in the form of state subsidies is not available (Muller 1966). No wonder some of the snow gauges operated by highway departments are placed in strategically located areas of maximum snowfall!

For hundreds of ski resorts in the Great Lakes area, lake-effect snow is an economic boon. The deep snowcover represents a resource in terms of available water to fill storage reservoirs and rivers. But floods are a danger during the spring thawing period, and structural damage is a constant concern. Muller states that in rural areas of New York snowbelts, ladders are left leaning against buildings during the winter to give easy access to roofs for removing accumulated snows. And on Michigan's Keweenaw Peninsula, residents construct elevated sidewalks which can easily be swept clear of snow (see photo, p. 173).

Is it possible to alter lake-effect snows in any way? Dr. Helmut Weickmann, director of the Atmospheric Physics and Chemistry Laboratory, NOAA, Boulder, Colorado, thinks so. He points out that the snow falling from a typical lake-effect snowstorm from Lake Erie contains about 250 billion gallons (1,390 billion liters) of water—enough for a 16-day supply of water for the entire population of 32 million within the Great Lakes basin. Thus the ability to control these storms may significantly affect human activities. Control may be feasible, says Dr. Weickmann, by seeding of lake-effect storms in order to redistribute the snowfall. In fact, it may be possible to dissipate some of the clouds themselves, thus alleviating the gloom which typifies the winters in the Great Lakes area (Weickmann 1972).

Rules of Thumb for Predicting Lake-Effect Snowfall

Suppose you reside within one of the Great Lake's snowbelts. What rules of thumb might you use to forecast the occurrence of a lake-effect snowstorm? Remember, these storms can occur anytime from October to March. They are most likely to occur during unusually cold periods, with

An elevated walk in the Copper Country of Michigan—a cultural
feature associated with heavy snowfall.

temperatures at freezing or much below and winds from the
west-southwest to northwest. The air over the lake water seems
to require at least a 50-mile trajectory for significant snowfall to
occur. Thus west winds are usually not heavy-snow winds over
Lake Michigan, but they may be over Lake Erie or Lake
Ontario. Over Lake Superior, a greater variety of wind directions may bring heavy snows. If you reside within a hilly
area, the chances of lake-effect snowfall are even better. And if
you have access to lake water temperatures, remember that the
greater the difference between water and air temperature, the
greater the chance of lake-effect snow.

   If all these conditions are met, it is a good bet that you will
need a snow shovel.

11: Winds and Waves

THE POWERFUL WINDS AND HIGH WAVES WHICH OCCUR from time to time on the Great Lakes have menaced shipping interests for decades. Besides shipwrecks, the winds and waves of the Lakes cause a continuing erosion of shorelines. During the high-water period of the early and middle 1970s, millions of dollars in damages were incurred along the shores of the Lakes because of wave and current erosion. Lakefront property was gradually eaten away by the force of the waves; entire cottages slid into the lake waters; and highways became undermined by retreating shorelines. Property owners, desperately attempting to preserve their land, resorted to various means of lessening the effect of the waves. Breakwalls of concrete and sand were hurriedly positioned, jetties and groins were erected, and emergency relief in the form of government loans and insurance payments was made available. While erosion becomes acute during high-water periods, it is also a problem during low-water stages.

How Do Waves Form and Grow?

Waves on the Great Lakes are caused essentially by the winds. Wave motion can best be described as *progressive*

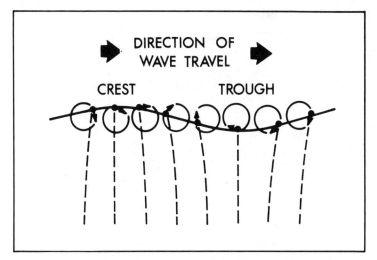

Figure 51   Orbital Motion in Deep-Water Waves of Low Height

oscillatory. In order to illustrate the meaning of these terms, let us trace the motion of a small particle within a wave field (figure 51). As the crests and troughs of the waves pass, the particles move through nearly circular orbits. However, they do not return to their exact points of origin but move slightly forward on each orbit. Thus, although most of the particle motion is oscillatory, a net forward movement also occurs. This forward movement is known as *mass transport,* and it is dependent on the height (vertical distance between successive troughs and ridges) and length (horizontal distance between successive crests) of the wave (figure 52). The rate of mass transport may reach 2 knots.

Waves grow by the push of the wind on the wave crest, thus driving the wave forward, and also by the *skin drag*—the frictional effect of air motion over water which tends to pull the water in the direction of the wind. The skin drag is greatest on the crests and least in the wave troughs.

The heights which waves may attain are related to several variables. One quite obviously is the velocity of the winds. Being a region of frequent cyclonic activity, the Great Lakes do

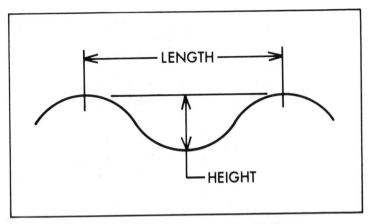

Figure 52   Wave Length and Height

not lack for strong winds. Another factor is the duration of the winds. A high-velocity wind which continues in a constant direction for many hours is a very effective wave producer— much more so than a sudden, violent squall, which may last only a brief period. For developing high waves, it appears that wind duration is most important during the early stages of wave development, although waves will continue to grow for periods up to 32 hours (Strahler 1961).

A third variable is fetch, the length of wind flow over the lake surface. The greater the fetch, the greater the potential for high waves. The map will reveal the importance of fetch for wave development on the Great Lakes. For example, maximum fetch would occur with northeast winds at Duluth, Minnesota; with north winds at Michigan City, Indiana; with west-southwest winds at Buffalo, New York; and with east winds at Toronto, Ontario. Therefore the highest waves might be expected with winds from these directions at these cities, other things being equal. The alignment of the various lakes would also tend to dictate which wind directions might generate the highest waves. On Lake Michigan, those would be north or south winds; while on Lake Erie and Lake Ontario, winds from the east or west would likely be the big wave producers.

How large may waves grow on the Great Lakes? This is a difficult question to answer. Not many accurate measurements exist, almost none for the middle of the Lakes.

Scientists at the University of Michigan have taken a different approach in attempting to determine both the range of wave heights and the maximum wave heights that might occur on the Great Lakes. Working from theory, they used records of wind fetch, wind duration, and mean wind speeds derived from past weather conditions in order to predict possible expected wave heights. This method is called *hindcasting*. The hindcasts indicated that, as expected, wave heights should have been greatest in winter and least during summer. On Lake Superior, with many storms and a minimum of fetch limitation, the hindcasting techniques indicated that a wind of 50 knots (91 kilometers per hour) for 12 hours duration would produce a wave height of 30 feet (about 9 meters). A wind of 60 knots (111 kilometers per hour) with a duration of 10.5 hours might be expected to produce a wave height of 33 feet (10 meters), and winds of 70 knots (130 kilometers per hour) and duration of 9.5 hours would produce a wave height of 50 feet or about 15 meters (Cole 1971).

Hindcasting suggested that maximum waves in excess of 56 feet (17 meters) would be expected each year in the middle of Lake Superior and that maximum waves of 69 and 77 feet (21 and 23 meters) could occur. How close are the values obtained by hindcasting methods to those actually occurring? It is difficult to say, but on Lake Superior, maximum waves of 69 and 77 feet (21 and 23 meters) may be close to reality.

Wave Erosion along Shorelines

Exactly how does a wind-driven wave train damage and erode a shoreline? Most of the energy of the wave is released when the wave breaks. Waves break upon approaching shallower water and, as a general rule, when the depth of water approaches 4/3 the height of the wave. In other words, a 3-foot wave would break upon approaching the shoreline when the depth of water became 4 feet or less. The breaking wave is caused by an enlargement of the water particle orbit to an elliptical form as the wave moves into shallower water. As the speed of wave movement decreases in shallower water, the velocity within the elliptical orbit may exceed the wave velocity and the wave will break.

Breaking waves loosen coastal materials, which can be transported back into the lake by the backwash of the wave. Another point to consider is that waves rarely approach a shoreline at right angles. Waves obliquely approaching shorelines set up *longshore currents,* which may transport and deposit materials at some point further along the beach. Thus during storm periods, both erosion and deposition may be occurring along a shoreline as the result of an almost random relation between meteorological variables and such factors as shoreline configuration, material types, and so on.

Man-made sea walls may be constructed in an attempt either to prevent waves from breaking or to absorb the energy release of the breaking wave. If the water is quite deep in front of a sea wall, the wave will not break and the energy will be deflected elsewhere. If a pocket of air is enclosed within the breaking waves, the air becomes compressed, and large shock pressures may occur when the air is released. The shock pressures are of short duration but do most damage along a coastline.

Jetties and groins may retard the transport of materials by longshore currents along the beach. They may effectively reduce erosion in the up-current beach areas, but very often the down-current beaches are excessively eroded because replacement materials are not made available.

Only a few storms during the year are responsible for most of the shoreline erosion, and these storms occur primarily in the fall and early winter. Up to 1959, it was reported (Strong and Bellaire 1965) that thirty-eight fall storms and only nine spring storms on the Lakes had caused serious damage.

November seems to be the month most prone to high wave occurrence. The reason for this, as for most of the meteorological and climatic effects caused by the Lakes, is related to the difference in temperature between the land and the water.

Studies have shown that during the stable season, the effect of cold air just above the lake surface and an inversion aloft produces a *blanket effect,* which retards the transfer of momentum from the wind to the water (Strong and Bellaire 1965). Thus smaller waves result. During the unstable season, however, when the air close to the lake is warmed, momentum is transferred very effectively. November is about the peak of the unstable season and is also the month when cyclonic storms start to become quite intense. Thus shipping interests and shoreline property owners must beware the month of November.

The build-up of ice along the shore during the winter is very critical in protecting the shore from wave erosion. Although the Great Lakes usually do not freeze over entirely, shoreline ice normally builds up and may extend some distance into the lake. The ice forms a protective rim around the lake, sparing the shore itself from the wave battering which would otherwise occur. The property owner welcomes the protective shore ice, as a mild winter with little ice may bring accelerated shoreline erosion.

Surges and Seiches

Suppose you have a shallow, elongated pan containing water. Now quickly lift the pan, then return it to its original position. Notice the behavior of the water. When the pan is lifted, the water accumulates at one end and the depth increases. When the pan is returned to its original position, the water sloshes back and forth, first deeper at one end, then at the other. A free oscillation of water continues for a short while, with the time between successive sloshes becoming shorter and shorter while the differences of water levels between opposite ends become smaller and smaller. Eventually the oscillation ceases and the water returns to its original even level.

Large lakes may behave in a similar fashion, particularly if the lakes are elongated and shallow. Free oscillations may occur on the lakes, and the water level of the lakes may change as the water sloshes back and forth. Such a phenomenon is called a *seiche* (pronounced "saych"). Seiches may occur on any of the Great Lakes but are particularly marked on shallow, elongated Lake Erie.

Of course, the analogy of the lake to the pan must be qualified. Obviously we do not lift up the lake basin at one end to start the oscillation. Something else must supply this energy to the lake waters. In certain cases, the energy available within the atmosphere can be transferred to the lake water. If persistent strong winds occur for a sufficient period of time, water is removed from the windward side of the lake and transported to the leeward side. The rate of transport is estimated to equal about 5% of the speed of the wind. On Lake Erie, for example, west-southwest winds have occasionally piled up water in Buffalo harbor several feet above the normal water level. Strong north winds over Lake Michigan or Lake Huron could accomplish the same thing. This pileup of water is known as a *setup* or *surge*. When the winds cease, the water returns through periods of free oscillation to its original state.

Sharp changes in barometric pressure existing in different portions of the Lakes can also cause surges of water which result in seiches. The lake surface lowers in areas of high pressure, and the water flows toward the area of lower pressure. In addition, squall lines moving across the Lakes may cause high winds which cause water surges to occur. Sudden jumps of pressure which accompany squall lines may also contribute to the piling up of water. With the passage of an intense low-pressure area across the Lakes, high winds, large pressure differences, and squall lines may coexist, thus setting up excellent potential conditions for a surge and seiche to occur.

Following the initial surge, the period of free oscillation of the water may vary according to the configuration and depth of the lake, its number of bays and inlets, the alignment of the channel axis, and other variables. Some theoretical studies (Rockwell 1966) have indicated that the predicted initial oscillation period on Lake Erie is 14.08 hours, with the second oscillation period predicted for 8.92 hours. Being a shallow lake, the oscillation periods of Lake Erie are longer than for the other Great Lakes. On Lake Michigan, the first mode of oscillation is predicted to be 8.83 hours, the second mode 4.87 hours. Oscillations on deep Lake Ontario have first modes of 4.91 hours and second modes of 2.97 hours. Observations of seiches on the Lakes have tended to support these theoretical calculations.

Seiches and surges, which cause changes of water levels along the shores of the Lakes, can endanger lives, affect water-intake facilities, cause changes in harbor depth, increase shoreline erosion, and damage dock facilities. Therefore whenever the probability for a seich is high, warnings are issued by the National Weather Service. When a seich warning is issued, it is best to avoid the Great Lakes if possible.

12: Effects on Hail, Thunderstorms, and Precipitation

THUNDERSTORMS OCCUR MORE FREQUENTLY IN THE southern portions of the Lakes, where temperatures are warmer. However, they may occur anywhere in the Great Lakes region during the summer. In winter they are very rare; in fact, the northern portions of the Lakes generally go from fall to spring without hearing the sound of thunder. On the average, thunderstorms occur on about 40 days in the western and southern portions of the Lakes basin, but on less than 25 days in the northern portion.

Thunderstorms require strong uplifts of warm, moist air. The uplift necessary for thundercloud formation may be caused by the presence of a front or an atmospheric disturbance which causes strong convergence of air near the surface. As mentioned in part 1, strong heating of the earth's surface may also play a role in developing thermals, or updrafts, which may have the potential of forming a thunderstorm cell. These uplift processes may occur simultaneously; that is, strong surface heating or convection may combine with an atmospheric disturbance to set off the occurrence of thunderstorms.

Local *hot spots* may be of considerable importance in instigating updrafts which eventually become thunderstorm cells. These hot spots are small areas of the earth's surface which may heat up quite strongly. A plowed field surrounded by forest may become a hot spot because it lacks vegetation. This lack allows more solar energy to reach the surface, and the plowed field also lacks the effective natural air conditioning provided by a plant or tree cover which transpires and evaporates moisture. Cities may become hot spots in the sense that they are usually warmer than the surrounding countryside. Even shopping centers or asphalt parking lots can occasionally become hot spots, and thermals or updrafts can then be detected.

The project METROMEX (METROpolitan Meteorological EXperiment) was a 5-year research effort in the St. Louis metropolitan area directed toward obtaining information on how cities modify weather and climate. We will discuss some of the findings of METROMEX and the implications to the climate of the Great Lakes area in a later chapter. For the purposes of this discussion the METROMEX findings show the importance of localized hot spots in the development of rain-shower cells. Areas were recognized within the boundaries of the city which either developed rain cells in the air overhead or where existing cells were intensified. As a result of MET-ROMEX and other recent studies, renewed emphasis has been given to the effect of local surface features in instigating (or suppressing) rain cells and thunderstorm cells.

How the Great Lakes May Affect Thunderstorms

The Great Lakes area contains some very large hot spots and also some very large *cold spots*. These are the Great Lakes themselves: hot spots during the unstable season and cold spots during the stable season. During the unstable season, the Lakes as heat sources give impetus to rising air and also contribute moisture to the atmosphere. The result is increased cloud cover, warmer temperatures, and more snowfall along the downwind shores. During the stable season, however, the Lakes are cold spots, suppressing vertical motion, causing inversions, and inhibiting cloud development; and they are characterized by the presence, on some days, of a mesoscale anticyclone. As thunderstorms need warm, moist air and rapid uplift for their development, it would certainly be logical to assume that the Great Lakes can play a role in determining the numbers of thunderstorms over and along the shores of the Lakes. But do they?

The staff of the Atmospheric Sciences Section of the Illinois State Water Survey was much concerned with this problem. Their job has been to evaluate the availability of water in the atmosphere to the state of Illinois. As Illinois borders on one of the Great Lakes (Lake Michigan), the Water Survey was concerned with the possible effects of Lake Michigan in either suppressing or enhancing precipitation. And because much of the summer rainfall around Lake Michigan results from thunderstorms, the role of the Lake in instigating or suppressing thunderstorm activity was also of interest. What was learned about Lake Michigan could also, with some adjustment for physical differences, be extrapolated to all of the Great Lakes.

A climatological study by Stanley A. Changnon, Jr. (1968), of the Illinois State Water Survey showed that Lake Michigan affected thunderstorms during all four seasons of the year but in different ways. As might be suspected, during the winter when the lake was warm relative to the land, the lake increased the occurrence of thunderstorms, even though they were relatively infrequent at that time of the year. In the spring, when the lake was colder than the land, the lake decreased thunderstorm occurrence along the western shore but increased them along the eastern shore. During the winter and spring, the effects of the lakes were minimal compared to fall and winter. In the fall, when the lake was warmer, it increased thunderstorm occurrence significantly along the eastern shore. But in the summer, when the lake was colder than the land and thun-

Flooded cornfield at Arlington Heights, Illinois, following 6¼ inches of rain.

derstorms are most frequent, the lake both increased and decreased thunderstorm activity. This bears further discussion: How can the lake both increase and decrease thunderstorm occurrence?

In our discussion of lake breezes, we compared the diurnal temperature curves of the lake versus the surrounding land as they might occur typically during the early summer. We found that at night the lake was usually warmer than the land, and that during the day it was cooler. Averaged out, the lake is cooler than the land. Results of the Illinois State Water Survey research found that during summer nights, the lake increased thunderstorm activity on both sides of the lake in the northern portions and along the east shore in southwestern Michigan. In northern portions, nocturnal cooling of the land was greater, thus creating a larger relative temperature difference at night.

In both the northern and southern portions of Lake Michigan, the lake reduced thunderstorm activity during the day in summer, when the lake was cooler than the land. But in the north, the increase in nocturnal thunderstorm occurrences offset the decrease in daytime occurrences so that the lake produced a net increase of thunderstorm activity during the summer. In the south, the suppression effect during the day offset the enhancement effect at night so that the lake's net effect was to decrease thunderstorm activity. Complicated? Yes, but the entire pattern results from the relative differences in temperature between the lake and the land.

How much effect does the lake exert in increasing or decreasing thunderstorms in summer? The Illinois State Water Survey concluded that the lake could cause as many as 20% more thunderstorm days during the summer in the northern part of the basin; while in the southern part of the basin, the number of thunderstorm days could be reduced by 20%. Annually, the effects during each individual season, some of which cancel each other out, must be considered. But the Water Survey study indicated that over the southern Lake Michigan basin, the lake causes an average decrease of 2 thunderstorm days per year along the eastern lakeshore; and in the northern part of the lake basin, 6 thunderstorm days per year occur along the eastern shore because of the effect of Lake Michigan.

Hail Occurrences and the Great Lakes

Hail is a weather phenomenon of considerable economic significance. Annual losses as a result of hail crop damage in the United States range close to $700 million and about 1.3% of total crop production is annually written off as hail loss (Changnon et al. 1977, p. 27). In addition, damages to structures, livestock, trees, and vehicles may occur. Although the Great Plains wheat areas and the corn belt of the Midwest lead in crop loss due to hail, the Great Lakes area is also vulnerable. During the summer of 1975, fruit losses in southwestern Michigan ran into millions of dollars because of hail damage.

Hail results from the growth of ice layers around an ice particle; this occurs within a cloud which extends far into a subfreezing environment. The ice particle grows through collisions with supercooled water droplets (water droplets which retain their liquid state even at below-freezing temperatures). The supercooled water droplets then freeze on the surface of the ice particle.

Certain processes must be operating in order for the ice particle to remain within the cloud for a very long time and thus collect the necessary moisture to grow into hailstone size. Perhaps the ice particle originated in the very high portions of the cloud. If the cloud is of great vertical thickness, the pellet might then accumulate enough ice in its descent through the cloud to grow into hailstone size. Another possibility: the pellet might be sustained by strong updrafts within the cloud, then be released when the hailstone has grown very large and heavy or has entered a part of the cloud where the updraft is not powerful enough to suspend it.

In either situation, it appears that the cloud must have considerable vertical thickness and abundant supercooled water in at least some of its portions before hailstones will form. The concentric layers of ice, opaque and clear, surrounding the pellet are thought to be caused by different concentrations of supercooled water which the pellet encounters during its residence time within the cloud.

A severe thunderstorm provides the optimum environment for hail to form. The cloud has both the great vertical thickness and strong updrafts which may be required for hail formation. If thunderstorm occurrence is influenced by the Great Lakes, wouldn't the Lakes operate either to enhance or suppress hail occurrences too? The Illinois State Water Survey was also interested in this problem and investigated the frequency of hail days around the shores of Lake Michigan.

The Water Survey (Changnon 1968) found that, during the summer, Lake Michigan caused a decrease in the number of hail days over much of Lower Michigan by as much as 10% to 60%. This suppressing effect results from the lake's role in lessening convective uplift as weather systems pass over. In the fall, however, the lake increases the number of hail days from 50% to 400%! (This figure seems a bit alarming, but remember that in the fall few hail days occur in the area anyway, and the hailstones are normally quite small. The Great Lakes region, however, is the only part of the United States with a fall hail maximum.) In the spring, at the height of the hail season, the lake has an inhibiting effect on hail occurrence, as might be suspected, because of its relatively cold temperatures. In winter, hail rarely occurs in the Great Lakes area because of the cold temperatures and lack of thunderstorm development.

The Illinois State Water Survey has estimated that lake effects cause an increase of as many as 8 hail days in 10 years, or a bit less than 1 day per year. For a farmer in southwestern Michigan, however, even 1 added hail day can be critical.

Lake Effects and Precipitation

We have already seen that the Great Lakes play a very strong role in determining the amount of winter snowfall. The lake-effect snowfalls, which develop when cold air passes over the warm lake water, cause from 25% to 100% more snowfall on the downwind shores of the Lakes.

We have also seen that the Lakes may suppress or enhance thunderstorms. The cited study by the Illinois State Water Survey showed that about one-half of Chicago's precipitation resulted from thunderstorms (Changnon 1968). Thus the thunderstorm is a significant cause of rainfall in the Great Lakes area.

Can an overall assessment be made about the role of the Great Lakes in either increasing or decreasing the total amount of precipitation? Changnon attempted to do so for Lake Michigan by studying long-term records for stations along the Lake Michigan shoreline, also by studying records available for a few islands in the lake, particularly in the northern portion. His study showed that the average annual precipitation over Lake Michigan is 6% less than over the surrounding land area. During the summer season, the average precipitation over the lake is 14% lower than over the surrounding land—a fact related to the effect of the lake in suppressing thunderstorms over much of its southern portion. His results differed from previous studies that concluded little or no difference existed between the lake and the land.

The reduction of precipitation over the lake by 6% doesn't sound like a lot, but it means that 700 billion gallons less water would fall over the lake. During high-water periods when the lakeshores are being eroded and cottages are destroyed, shoreline dwellers may be thankful that the Lakes suppress precipitation. An extra 6% could well mean complete disaster. However, during low-water periods an additional 6% to maintain navigation channels, harbor depths, and adequate freshwater supplies might well be desirable. Is there any possibility of altering the amount of precipitation received over the Lakes? We will attempt to answer this question in a following chapter.

Not only are precipitation amounts altered over Lake Michigan itself but also along the shorelines. The Illinois State Water Survey study showed that the amount of thunderstorm rainfall is lower on the east side than on the west side of the lake, at least for the southern lake basin. Part of this difference is due to a natural decrease in thunderstorm activity from west to east over the basin and would exist even if Lake Michigan was not present. Part, however, is due to the effect of the lake in suppressing thunderstorm activity. In the northern portion of the basin where the lake tends to increase the number of thunderstorms, the relationship is likely to be reversed.

Although the work of the Illinois State Water Survey has told us a great deal about how the lake may interact with the atmosphere to suppress and enhance thunderstorms and precipitation around Lake Michigan, the studies completed have also warned us that the relationships are very complex. Regional differences exist, for example, in summer thunderstorm activity as it is suppressed in the southern part of the lake and increased in the northern part. To make a single conclusive statement on the role of the lake is difficult because in some cases opposite effects are occurring at the same time. Thus we must use caution in extrapolating the Lake Michigan findings to the other Great Lakes. We can state with some certainty, however, that all the Lakes affect the occurrences of thunderstorms, hail, and the amounts of precipitation over them and along their shores. How, when, and how much are questions which cannot presently be answered with a high degree of certainty.

PART THREE

Climatic Patterns in the Great Lakes

13: The Climate
of the Great Lakes

WHILE THE TERM 'WEATHER' REFERS TO THE COMBINA-
tion of physical effects occurring within the atmosphere at a
given time and place, *climate* is the composite of weather over a
long period. In a large way, climate determines how people
live, including how they earn a living, what they do for
recreation, how they dress, and the way they feel. The
agricultural activities of an area and even its industrial base are
closely related to climate. Although modern conveniences such
as interior heating, air conditioning, the automobile, and
innumerable energy-expending devices allow some escape
from the firm grasp of weather and climate, we are never
entirely free from its hold.

Indeed, climate may give unity to a region, and regional
distinctiveness is often based upon some aspect of climate. We
think of southern California within the context of bright
sunshine and mild temperatures, of Florida as a mecca for sun
worshipers, or of the Pacific Northwest as shrouded by cloud
and rain.

The Concept of Climatic Classification

Areas which have uniform climatic characteristics are
known as *climatic regions*. An assembly of climatic regions and

the recognition of identifiable climatic features which distinguish each region creates a *climatic classification*. The goal of climate classification is the recognition of large areas having rather similar climatic characteristics so that the burden of describing the world's climates may be eased by grouping these somewhat similar areas together.

Attempts to classify the world's climates are not new. The ancient Greeks formulated a rather simple classification by recognizing three zones in each hemisphere which were differentiated by temperature. A torrid zone, where it was always warm or hot, existed near the equator. Near the polar regions, a frigid zone existed which was always cold. Between the equator and the poles was the temperate zone which was seasonal in nature, sometimes warm and sometimes cold. The Greeks resided within the temperate zone and experienced warm summers and cool winters.

One shortcoming of this very early climatic classification was that it failed to consider differences in precipitation which could exist within each of the climatic regions. For example, both dry and wet regions could exist within the torrid zone, and the Greek classification could not take these contrasts into account. Consequently the more modern and sophisticated efforts at climatic classification have usually been based on both temperature and some aspect of moisture availability.

During the past 100 years, numerous climatic classifications have appeared, and they continue to appear. They range from very straightforward and descriptive to elegantly devised classifications utilizing the assistance of computers for their formulation. None is perfect, but all have the similar goal of recognizing areas of the world which have similar climatic features.

Classification of the Great Lakes area according to the Köppen scheme

One of the most frequently used classifications is that of Dr.

Wladimir Köppen of the University of Graz, Austria. Köppen first devised his classification in the late 1800s. He felt that the vegetation of an area was the best indicator of the type of climate which the area experienced. Thus his original classification was based upon vegetation types and he gave the climatic regions vegetation names. Subsequently he dropped the vegetation nomenclature from his classification and took on a symbolic scheme in which the meaning of the symbols indicated the annual and monthly means of temperature and precipitation.

Köppen's classification was first published in 1918 and has gone through numerous revisions. A map of the world showing the location of the regions defined by his classifications shows that most of the Great Lakes region of North America has a climate shown by the symbols *Dfb*. Köppen called this climatic type *humid continental cool summer*. The fact that most of the Great Lakes basin is similarly classified doesn't mean that climatic differences do not exist within the region; but simply that some broad, unifying similarities within the area exist which justify the inclusion of the region within one climatic type.

Köppen gave the following explanation for his symbolic designation *Dfb:* The *D* refers to cold and snowy forest climates, with average temperatures of the coldest month below $-3°C$ (26.6°F) and average temperatures of the warmest month above $10°C$ (50°F); the *f* symbol implies that there is no dry season during the year; and the *b* symbol indicates that summers are cool, with the average temperature for the warmest month not exceeding $22°C$ (71.6°F). Only limited areas in the extreme northern and southern portions of the basin are not classified by Köppen as Dfb.

The map (figure 53) shows other areas of the world sharing the same Köppen designation as the Great Lakes. Note that this type of climate is confined to the Northern Hemisphere. The Southern Hemisphere lacks the large land masses within the middle latitudes necessary for the existence of the Dfb climate.

Figure 53   Köppen Dfb Regions of the World

Rather limited parts of North America and Eurasia share this climatic type; and other than the northern United States and southern Canada, the most extensive occurrence of Dfb climate is in the European portions of the Soviet Union. What really makes the North American section unique, however, is the existence of large water bodies within the region. Nowhere else does this occur.

Köppen's symbols rather appropriately sum up the unifying features of the climate of the Great Lakes region. It is an area characterized by cold, snowy winters, cool or moderately warm summers, and evenly distributed precipitation. To these common characteristics we should add that variability is also a climatic keynote within the region. Large variations from the average weather may occur from time to time as a response to a given weather-map (synoptic) pattern. Additionally, considerable variation of individual weather elements has been recorded within the region. Temperatures have been as hot as 112°F (44.4°C) and as cold as −61°F (−46.1°C), and yearly rainfall totals have ranged from 64 inches (1,625 millimeters) to as little as 7.76 inches (197 millimeters).

While the term "climate" implies an "average weather" over long periods of time, we must emphasize that actual weather is rarely "average" and that significant departures from the average are particularly likely to occur within the Great Lakes basin.

The Climatic Controls

Four major controls exert a marked influence over the climate of the Great Lakes area. These are: (1) latitude; (2) air masses and atmospheric disturbances; (3) the continentality of the region (related to its position within the interior of North America); and (4) the modifying marine effects resulting from the presence of the Great Lakes.

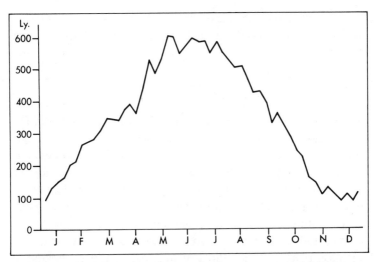

Figure 54   Annual March of Radiation at East Lansing, Michigan (values are 50% probabilities of mean daily values for weeks during the year)

Latitude

Latitude is the dominant control, insuring that large differences in solar radiation will occur seasonally. The annual march of solar radiation (in langleys) is shown in figure 54 for East Lansing, Michigan. The marked contrast in energy received during the winter as compared to summer means that sharp seasonal temperature contrasts will occur within the region. The strong contrasts between seasons of the year probably form the most salient climatic feature of the Great Lakes. Residents take the seasonal progression of weather for granted, rarely considering the fact that large portions of the earth's surface do not experience such substantial differences between winter and summer weather as does the Great Lakes area. Social, political, and economic structures in the Great Lakes area are, however, consciously and subconsciously tuned to the yearly progression of spring, summer, fall, and winter. What changes in life styles would occur if marked seasonal change no longer existed in the Great Lakes area!

Air masses and atmospheric disturbances

Related to its latitudinal position but also as a result of the Great Lakes location within the broad lowland corridor extending from the Arctic to the Gulf of Mexico, the region experiences rapid air mass exchange and a high frequency of atmospheric disturbances. As mentioned in part 1, the jet stream is frequently over the area, and its presence is associated with mobile atmospheric disturbances including cyclones, fronts, and anticyclones. Table 2 shows the mean latitudinal location of the jet axis (area of maximum velocity winds) at about 18,000 feet over the eastern United States and Canada for the months of November through August. The data show that the jet is located south of the Great Lakes throughout much of the winter season but moves quickly northward during April and May to take a position over central Canada during the summer.

The changing seasonal position and intensity of the jet stream control the seasonal frequencies of cyclones and anticyclones. Figure 55 shows the frequency of these figures tabulated for four portions of the Great Lakes basin over a 40-year period. Although differences occur between the subdivisions of the region, it is apparent that the frequency of cyclones increases during the winter when the jet swings southward over the eastern United States; and that the frequency of anticyclones is greatest in the summer when the jet is weaker and over central Canada. It is also apparent that the frequency of cyclones is greatest in the northern part of the basin.

Accordingly, the seasonal frequencies of air mass types also reflect the seasonal change in jet stream position. Studies of air

Table 2

LOCATION AND SPEED
OF AVERAGE MAXIMUM WESTERLY
WIND COMPONENT
500 MB. SURFACE

Month	Location of center of closed isotach Latitude (N)	Longitude (W)	Speed over Great Lakes (meters per second)	Direction over Great Lakes
November	39°	84°	22	270°
December	39°	74°	26	300°
January	42°	72°	28	280°
February	37°	75°	26	300°
March	35°	75°	24	290°
April	40°	65°	20	290°
May	44°	64°	16	290°
June	47°	70°	——	——
August	50°	50°	——	——

The data indicate the core of maximum west winds (jet stream) in relation to the Great Lakes. Data from Lahey, Bryson, Wahl, and Henderson, *Atlas of 500 mb Wind Characteristics for the Northern Hemisphere.*

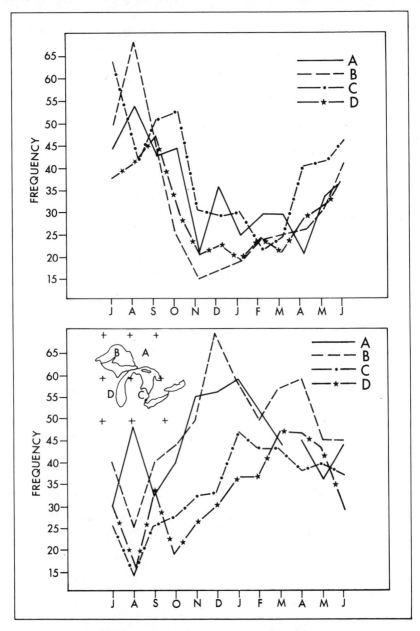

Figure 55 Frequency of Anticyclones *(top)* and Cyclones *(bottom)* for Subdivisions of the Great Lakes Area over a 40-Year Period

mass frequencies at Toronto in the winter (Smith 1971) have shown that about three-fourths of the days experience air masses of Pacific origin, highly modified after long passage across the United States and Canada. The remaining days are dominated by very cold polar continental air which has originated in Arctic regions. In the winter, maritime tropical air from the Gulf of Mexico rarely reaches as far north as the Great Lakes basin, occurring on less than 3% of the days at Toronto.

In the summer, air masses from the Pacific and Gulf of Mexico occur with about equal frequency, each on about 30% to 40% of the days (Bryson 1966). The remaining days experience cool and dry air masses from northern Canada and hot and dry air masses from the southwest deserts.

Continentality

The continentality of the region is related to its position in the interior of North America, where it responds to the more rapid heating and cooling rates of the large land mass surrounding it. Continentality is not readily translated into distinguishable climatic values but involves a host of climatic features normally associated with large land masses rather than with large water bodies. A convenient mode of expression of continentality may be obtained by comparing the differences of mean temperatures between winter and summer for stations of similar latitudes or for stations where the effect of latitude in causing seasonal contrasts has been removed. The essential idea is that where areas are primarily controlled by the quick-heating response of large land masses, large temperature differences will occur between winter and summer. Where large water bodies with their slower heating and cooling rates are the more dominant control, smaller differences between winter and summer temperatures will occur.

A formula which attempts to measure the role of continentality in governing the climatic character of an area has been

devised by climatologist Victor Conrad.* His formula elimi-
nates latitude as a factor in causing the annual range of
temperature, and the resulting index value ranges from 0 for the
least continental (most marine) station to 100 for the most
continental (least marine) area. While admittedly not a perfect
expression of the concept of continentality, it does allow
comparisons to be made from place to place, regardless of
latitude, of the amount of influence exerted by either continental or marine factors.

A map of continentality values for the Great Lakes area
(figure 56) shows that the climate of the region is primarily
continental with large annual ranges of temperature. The
presence of the Great Lakes, however, causes a noticeable
lessening of continentality along the shores and a general
reduction of the continentality which would be present were it
not for the Lakes.

Effect of the Great Lakes

The marine effects imposed by the Great Lakes constitute
the fourth major control of climate. As described in part 2, these
effects operate in different ways at different times of the year
and are felt to varying degrees around the basin. No part of the
basin escapes these effects, however.

Table 3 summarizes some estimated effects over, and
downwind from, Lake Michigan due to the moderating effect of
the lake in winter and summer. Similar modifications of
climate would be expected to occur along the shores of the other
Great Lakes, although the estimates of quantitative values
might differ.

*The mathematical derivation is as follows:

$$K = \left[\frac{1.7A}{\sin(\phi + 10)} - 14 \right]$$

where K = continentality index, A = annual range of temperature (in degrees
Centigrade, ϕ = latitude (Conrad and Pollack 1950).

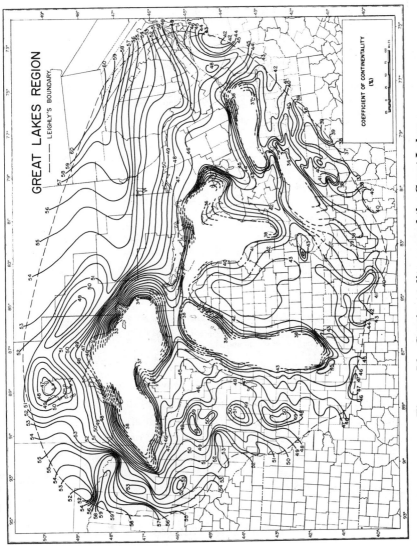

Figure 56 Continentality around the Great Lakes

Table 3
THE LAKE EFFECT

Variations ascribable to the effect
of Lake Michigan on weather phenomena
over and downwind of the lake (in percentages).

Weather Condition	Summer (June–August)		Winter (Dec.–Feb.)	
	Midlake	Downwind	Midlake	Downwind
Cloudy Days	−10	−15	+20	+35
Solar Radiation	+21	?	−10	?
Percent Possible Sunshine	>+5	+ 5	−10	−30
Snowfall	—	—	+ 3	+50
Precipitation	−15	−10	+ 7	+25
Days with ≥ 0.01 inches	− 7	−11	+17	+45
Thunderstorm Days	± 5	±10	+10 (fall)	+25 (fall)
Hail Days	−15	−33	+25 (fall)	+100 (fall)
Surface Wind Speed	+30	± 5	+98	+11
Mean Maximum Temp.	− 9	− 3	+ 3	+ 6
Days with > 90°F	−37	−50	—	—
Mean Minimum Temp.	± 1	− 2	+10	+15
Days with > 32°	—	—	− 3	− 6
Days of Heavy Fog	+70	+50	−16	−28

MODIFIED from Stanley A. Changnon, Jr., and Douglas M. A. Jones, "Review of
the Influences of the Great Lakes on Weather," *Water Resources Research*, 8, no. 2,
table 1, p. 369. Values are observed and estimated.

Stable season (summer) modifications by the Lakes are
most marked along the downwind shore with increases in heavy
fog days (50%) and hail (33%) and decreases in the number of
days with temperatures equal to or greater than 90° (50%).
During the unstable (winter) season, the most profound climatic
effects due to the Lakes are increases in cloudy days (35%),
snowfall (50%), days with .01-inch precipitation (45%), and
hail days (+100%).

Temperature Patterns

The role of the Great Lakes in modifying the temperatures
of the surrounding shores is one of their most fundamental

effects on the climate of the area. Economically, the spring cooling exerted by the Lakes prevents premature budding of sensitive fruit trees, lessening the chances of crop loss due to late spring frosts. The slow cooling of the Lakes in the autumn retards the occurrence of the first fall frost, thus extending the growing season. The length of the growing season (frost-free period) is thus related to proximity to the Lakes (figure 57).

Temperature patterns respond closely to the four climatic controls outlined. Latitude dictates that summers will be much warmer than winters and that the southern part of the basin will be warmer than the northern part. The daily weather-map pattern controls the temperature distribution over the region at any given time. Warm spells may typically accompany the presence of certain synoptic types (figure 58), while cool or cold spells are more common with others (figure 59). The continentality of the region insures that a large range of temperature will exist seasonally; and the lake effects cause cooler temperatures to occur along the shores of the Lakes in summer, and warmer temperatures in winter.

Winter and summer temperatures in the Great Lakes region

Various averaged values are commonly used to express the climatology of temperature. These statistics, in turn, can tell us much about the climate of any location or area, although mean or average values certainly do not tell the whole story.

The basic averaged temperature value is the *mean daily temperature*, and other statistics are derived from this. Mean daily temperature is computed by summing the maximum and minimum temperatures which occurred during the day and dividing by two. The maximum and minimum temperatures are recorded by special thermometers housed in standard instrument shelters within a dense network of stations in both the United States and Canada. Sum of daily mean temperatures

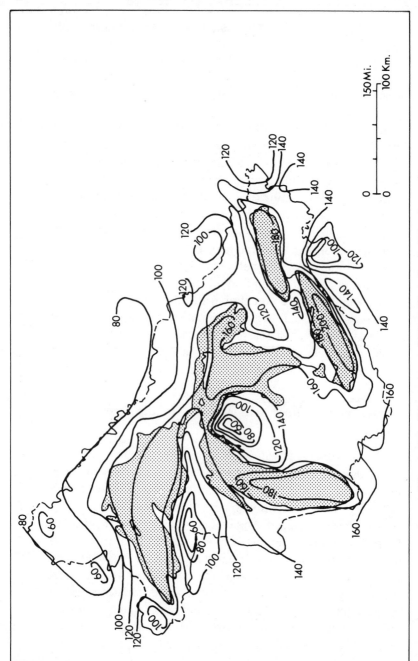

Figure 57   Mean Annual Frost-free Period (in days)

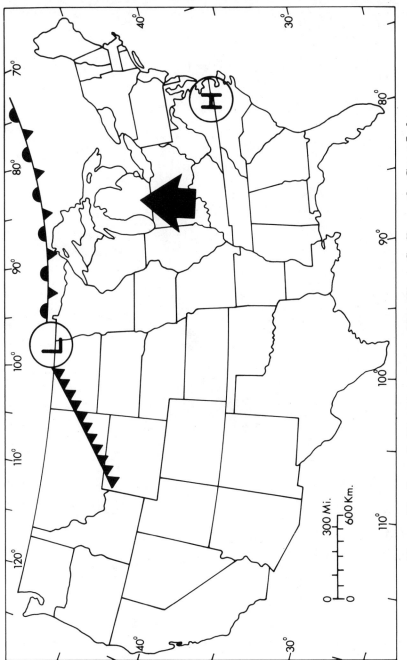

Figure 58   Synoptic Type Associated with a Warm Spell over the Great Lakes

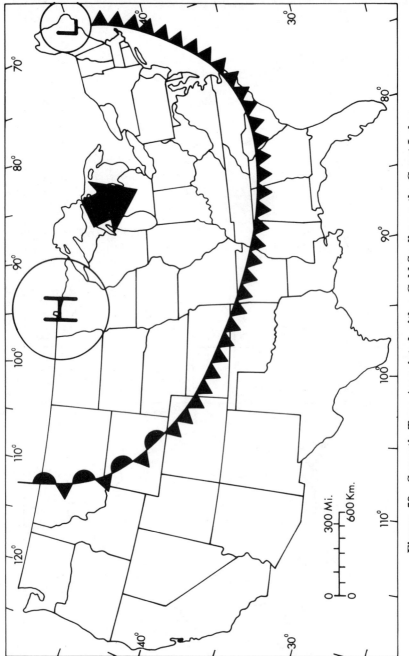

Figure 59 Synoptic Type Associated with a Cold Spell over the Great Lakes

during the month, divided by number of days in the month, gives the *mean monthly temperature;* and the sum of mean monthly temperatures, divided by twelve, is the *mean annual temperature.*

Mean monthly and annual temperatures, although useful to the climatologist, may be less meaningful to the layman because such statistics are difficult to equate on an everyday basis. Similarly, mean daily temperatures may mask a rather large range of values which may be encountered during the 24-hour period and thus may lose meaning to the individual. People in all walks of life, however, are accustomed to hearing morning weather predictions of the maximum temperatures for the day, and their sensitivities are attuned to the temperatures expected to exist during the working day. Therefore *mean daily maximum temperatures* for various months of the year (the sum of the daily high temperatures divided by the number of days) are statistics which may relate more directly to the average individual's daily experiences.

Figures 60 and 61 show the mean daily maximum temperatures which may be expected to occur within the Great Lakes basin for the months of July and January. The temperatures are shown by means of *isotherms,* or lines connecting points of equal temperature. When the isotherms are close together, the temperature gradient is sharp; that is, the expected afternoon temperatures vary within a short distance. Thus the isotherm pattern can be used to convey much information quickly about the temperatures over the area.

In July, normal afternoon temperatures range from 86°F (30°C) over the extreme southern part of the basin to 74°F (23°C) in the extreme northern portions—a variation of 12°F (6.6°C). The decrease in afternoon temperatures from south to north is not even, however, as temperatures over and along the shores of the Lakes are cooler. Lake Superior, largest and deepest of the Lakes, has the most marked effect on afternoon

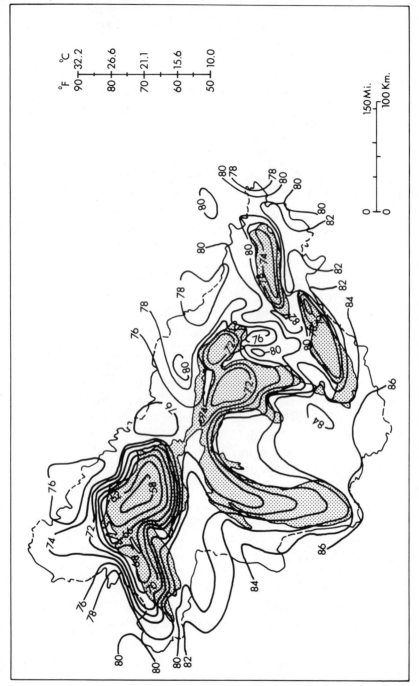

Figure 60   July Mean Daily Maximum Temperature

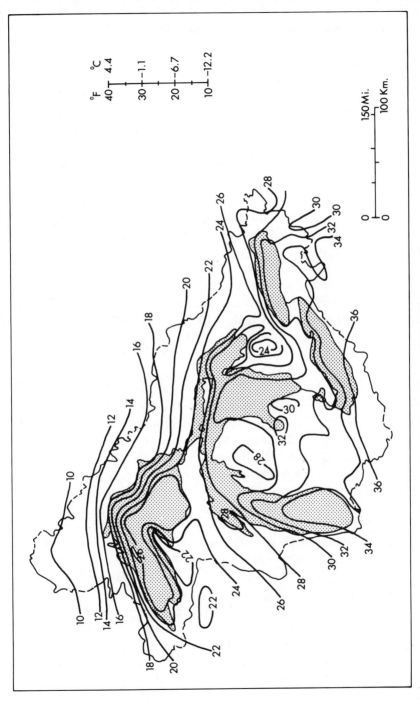

Figure 61   January Mean Daily Maximum Temperature

temperatures, with mean values failing to rise above 60°F (15.5°C) over a significant portion of the lake, and with a temperature range of 18°F (10°C) between overlake and interior locations.

In January, the latitudinal gradient is steeper, with afternoon temperatures averaging above 36°F (2°C) in the extreme southern parts of the basin and less than 10°F (−12°C) in the extreme north—a range of 26°F (14.4°C). The effect of the Lakes in altering the temperature pattern is less conspicuous than in the summer, although Lake Superior again exerts the strongest effect—but with a variation between overlake and interior locations of 8°F (4.4°C), as compared to 18°F (10°C) in summer. In winter, however, the effects of the Lakes are stronger with regard to mean daily minimum temperatures (maps not shown). In general, the Lakes most markedly affect afternoon temperatures in the summer and nighttime temperatures in the winter.

Precipitation

The term *precipitation* includes rain, snow, sleet, freezing rain, hail, and any other aqueous deposit received from the atmosphere—although over the Great Lakes basin, precipitation is received mostly as rain or snow. Precipitation in the Great Lakes area is abundant but not excessive. In the winter, most precipitation over the northern portion of the basin is in the form of snow, while in the south both rain and snow may occur. The liquid equivalent of the snow (the amount of water after the snow is melted down) is included in the annual precipitation total, given in inches or millimeters. Measurements of precipitation are made in the United States by gauges with a collecting diameter 31 inches (78.7 centimeters) above the ground and 8 inches (20.3 centimeters) in width. In Canada, the standard rain gauge stands 10 inches (25.4 centimeters) high and has a collecting diameter of 3.57 inches (9 centimeters).

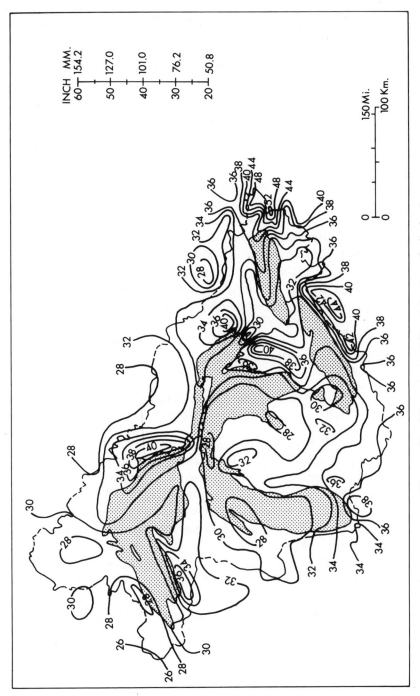

Figure 62 Mean Annual Precipitation (in inches)

On a global basis, the mean annual precipitation ranges from near zero in some extreme desert areas along the coast of Chile to 472 inches (11,739 millimeters) on the slopes of Mount Waialeale, Kauai, Hawaii. In the Great Lakes, the range is much less, from around 28 inches (711 millimeters) annually in the extreme northwest part of the basin to over 50 inches (1,270 millimeters) in northern New York State (figure 62). Precipitation increases southeastward across the Lakes, although the increase is not even, due to topography and lake effects. It is interesting to note that the evenness of monthly precipitation is also more pronounced in the eastern part of the basin. Graphs showing monthly rainfall totals for the Northeastern Climatological Division of Minnesota and the Great Lakes Climatological Division of New York State (figure 63) show that in Minnesota the precipitation is more strongly concentrated during the summer months, reflecting the area's greater continentality. With increasing continentality toward the western portion of the region, the number of days on which precipitation is recorded also decreases.

How Comfortable Is the Climate of the Great Lakes?

While it is customary for climatologists to describe the climate by statistics derived from individual weather elements such as temperature, precipitation, pressure, or humidity, these statistics certainly do not tell the whole story. Weather involves a combination of individual elements at a given time and place, while climate implies the frequency of occurrence of these unique combinations over a long period of time. It is the combined effect of the individual weather elements that people feel and respond to, but this combined effect is very difficult to measure.

For example, during periods of cold weather people must

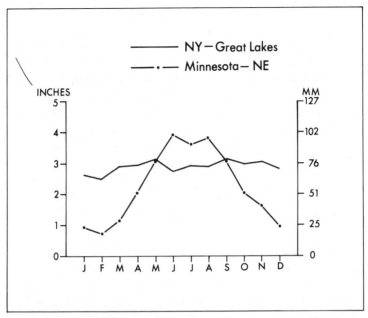

**Figure 63   Mean Monthly Precipitation Amounts, Northeastern
Climatological Division, Minnesota, and Great Lakes
Climatological Division, New York**

adjust to the cold stress to which they are subjected. If more energy is being removed from the body than is being received, the deep body temperature will lower, resulting in coma and eventual death. Adjustments, both voluntary and involuntary, must be made to prevent the lowering of body temperature. Some voluntary adjustments which can be made are obvious. The individual can leave the cold-stress environment (go indoors), or wear additional clothing which will prevent the loss of heat from the body. Involuntary physiological adjustments include shivering (to increase the metabolic rate and the production of internal energy) and constriction of the blood capillaries (to reduce the transfer of heat from the deep body to the extremities). Similarly, voluntary and involuntary adjustments occur when individuals are subjected to heat stress.

The amount of cold or heat stress does not depend solely upon external temperature but on a combination of temperature with other weather elements. For example, the amount of cold stress to which an individual is subjected depends on temperature, wind velocity (wind removes heat from the body by convective transport), amount of moisture in the air, and amount of radiation being received on the body. Measurement of the combined effect of all these weather elements is complicated, and the exact degree of cold stress is further complicated by the individual's physical condition, dimensions, mental outlook, amount of activity, and, of course, the amount and type of clothing.

A rather simplified but convenient expression of cold stress can be obtained from the *wind chill index,* which combines temperature and wind velocities to give a rough index of the amount of cold stress to which a normal individual would be subjected (table 4). Similarly, the *temperature-humidity index (THI)* measures the approximate heat stress to which a normal individual might be subjected when both temperature and moisture are considered (figure 64).

Somewhere between cold stress and heat stress is the range of conditions which seem to be comfortable for most people. Within this range lie the most comfortable climates, which are difficult to define objectively because comfort means different things to different people. One thing is certain, however: comfort involves a combination of weather elements—not just single weather elements—and comfort is an extremely important but often poorly described or considered aspect of climate. The millions of inhabitants of the Great Lakes area are directly affected by the comfort or lack of comfort which their climatic environment provides. How often is the weather comfortable in the Great Lakes area? How often are individuals subjected to climatic stress, whether heat or cold?

An approach to the mapping of comfort conditions within

Table 4

WIND CHILL CHART

Actual Thermometer Reading (in degrees Fahrenheit)

Estimated Wind Speed MPH	50	40	30	20	10	0	−10	−20	−30	−40	−50	−60
	Equivalent Temperature (in degrees Fahrenheit)											
Calm	50	40	30	20	10	0	−10	−20	−30	−40	−50	−60
5	48	37	27	16	6	−5	−15	−26	−36	−47	−57	−68
10	40	28	16	4	−9	−21	−33	−46	−58	−70	−83	−95
15	36	22	9	−5	−18	−36	−45	−58	−72	−85	−99	−112
20	32	18	4	−10	−25	−39	−53	−67	−82	−96	−110	−124
25	50	16	0	−15	−29	−44	−59	−74	−88	−104	−118	−133
30	28	13	−2	−18	−33	−48	−63	−79	−94	−109	−125	−140
35	27	11	−4	−20	−35	−49	−67	−82	−98	−113	−129	−145
40	26	10	−6	−21	−37	−53	−69	−85	−100	−116	−132	−148
Wind speeds greater than 40 MPH have little additional effect	Little Danger for Properly Clothed Person			Increasing Danger			Great Danger					
				Danger from Freezing of Exposed Flesh								

To use the chart, find the estimated or actual wind speed in the left-hand column and the actual temperature in degrees Fahrenheit in the top row. The equivalent temperature is found where these two intersect. For example, with a wind speed of 10 mph and a temperature of −10°F the equivalent temperature is −33°F. This lies within the zone of increasing danger of frostbite, and protective measures should be taken.

the United States has been made by bioclimatologist Werner Terjung (1966). Using a comfort index based upon varying combinations of temperature and humidity, Terjung mapped six categories of human comfort response over the United States for July and for January (figures 65 and 66). These categories attempt to describe the normal individual's reaction, but they are, of course, subject to widely varying individual responses because of varying physical and psychological make-ups. The maps point out that the average individual's response to the daytime climate of the Great Lakes region in July lies in the warm category. Along the immediate shores of

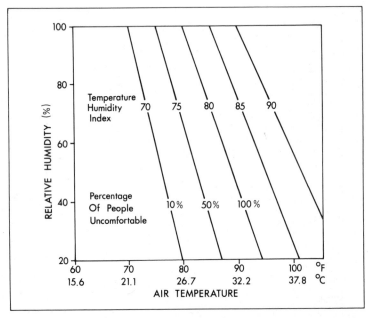

Figure 64   Temperature-Humidity Index. Read horizontally from the relative humidity, and vertically from the air temperature. The point of intersection gives TH index and the percentage of persons who may feel discomfort.

the upper Lakes, the comfortable category exists. Note the steep comfort gradient which exists just to the south and west of the Great Lakes; within 100 miles, one passes from warm to hot to oppressive daytime comfort conditions.

In January, as expected, conditions range on the cold side of the comfortable category—from cold in the northern part of the basin to keen in the southern part. In short, the Great Lakes daytime climate is a bit too warm in the middle of summer,

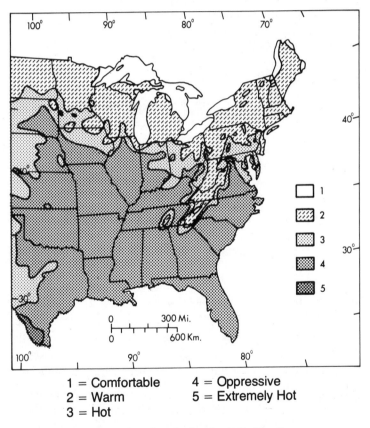

1 = Comfortable   4 = Oppressive
2 = Warm   5 = Extremely Hot
3 = Hot

Figure 65   Comfort Index for July Daytime

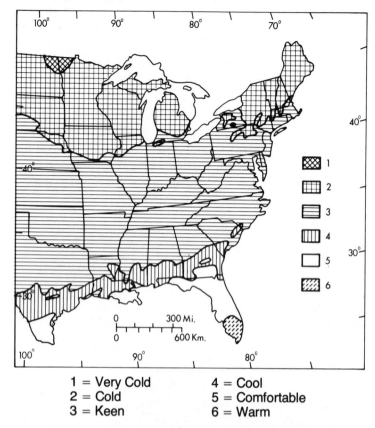

1 = Very Cold   4 = Cool
2 = Cold   5 = Comfortable
3 = Keen   6 = Warm

Figure 66   Comfort Index for January Daytime

except along the immediate shorelines of the upper Lakes, and is considerably too cold for comfort in the winter.

Maps of mean comfort conditions in July and January mask the availability of comfortable weather during the transition seasons, which many people insist are the most pleasant times of the year in the Great Lakes area. At the same time, the strong, short-term, day-to-day variability of comfort classes is not portrayed. This variability of comfort classes for January in Grand Rapids, Michigan, is shown by figure 67. Cold conditions occur for more than two-thirds of the month (66.9%), while keen and very cold conditions occur for 18.5% and 13.7%

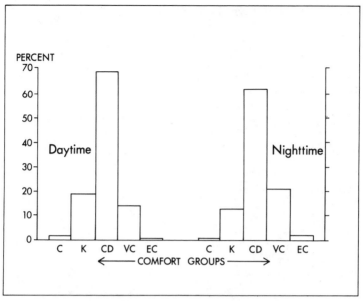

C = Cool   VC = Very Cold
K = Keen   EC = Extremely Cold
CD = Cold

**Figure 67   Frequency of January Comfort Groups
for Grand Rapids, Michigan**

of the month. Figure 68 reveals the spell-like comfort climates
of Grand Rapids for individual Januaries, showing the dominance of cold comfort climates but also the occurrence of keen
and very cold spells.

No single comfort class is as dominant in July as cold is in
January (figure 69). The comfortable (mild) and warm
categories occur for 31% and 42.1% of the month, with keen,
cool, hot, and sultry conditions occurring during the remainder.
Spell-like comfort categories are evident (figure 70), as in
winter, although spells tend to be more short-lived.

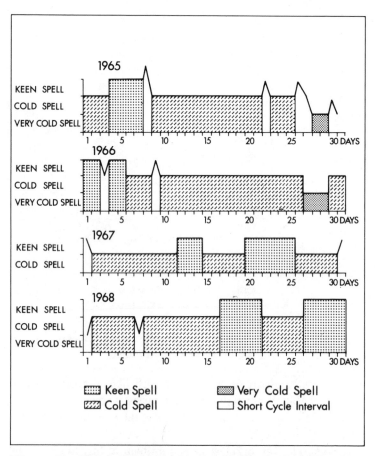

Figure 68 January Monthly Comfort Profiles at Grand Rapids, Michigan. The spells are periods of 2 or more consecutive days. During keen periods, temperatures of approximately 35° F are exceeded during the daytime extreme. During cold spells, the temperature at the daytime comfort extreme ranges from 14° F to about 35° F, and during very cold spells, temperatures do not exceed 13° F.

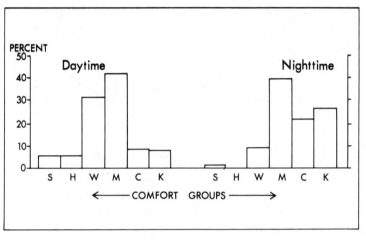

S = Sultry   M = Mild
H = Hot   C = Cool
W = Warm   K = Keen

**Figure 69   Frequency of July Comfort Groups
for Grand Rapids, Michigan**

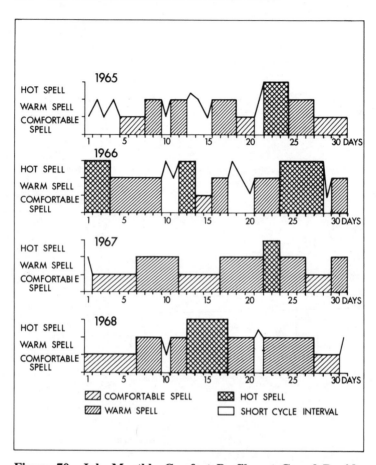

Figure 70   July Monthly Comfort Profiles at Grand Rapids, Michigan. These are spells of 2 or more days duration. During a hot spell, the average individual may experience considerable heat discomfort. During a warm spell, a relatively minor degree of discomfort may be experienced, and during a comfortable spell, the average individual experiences no heat discomfort.

14: Climatic Change in the Great Lakes Region

IS THE CLIMATE CHANGING? HOW OFTEN THIS QUES-
tion is asked. Various assertions regarding the climate of the
"old days" are readily forthcoming from grandmother, from the
farmer down the lane, and from the fisherman readying his
tackle and bait. During recent years, the news media have
echoed the weatherman in proclaiming that the global climate is
in transition. We hear of drought in the Sahel and India, of
unprecedented early fall and late spring frosts in the midwestern
United States, and of unusual dryness in southern California
and the Great Plains. And the unparalleled severity of the
winter of 1976–77 in the Great Lakes area stirs additional
suspicions of climates in transition.

Although it has long been known that, on a geological time
scale, past climate has been much different from that of the
present, only recently has it been realized that significant
climatic fluctuations have occurred during historical times.
Several events of an economic nature which occurred during the
early 1970s have focused attention on climatic variability.

After many years of substantial grain surplus, the world in
1972 faced diminishing grain supplies. The ready grain surplus
suddenly became less than the amount needed to offset a

225

weather disaster. In addition, the energy crisis, which was many years in the making but first became apparent as a result of the Arab oil embargo of 1973, is also climate related. Heating fuel requirements bear a close relation to the severity of the winter season, and the warmth of the summer determines the demand for air conditioning.

As a result of the role of weather variation in either sharpening or mitigating these economic problems, the capability of recognizing past climatic variations and of predicting future vacillations took on a new note of urgency. Climate suddenly became a critical factor in the capability of the world to feed its people; for the first time, national and international decision making became sharply responsive to climatic conditions.

Over 30 million people reside within the Great Lakes basin. It is a region of substantial economic significance, with agricultural, manufacturing, and recreational industries of great importance both in the United States and Canada. The question of climatic change and weather variation is one of utmost pertinence.

Past Climates of the Great Lakes

Changes of climate over the geologic time scale

Abundant evidence shows that past climates in the Great Lakes area were substantially different. On a geologic time scale, the changes were particularly well marked and the evidence conclusive. The sedimentary rocks underlying the Great Lakes basin give ample proof of warm and dry climates that existed millions of years ago. Salt and gypsum deposits attest to high evaporation rates and aridity existing at various times in the past, conditions which are not present today. Coal deposits suggest past warm and humid conditions with an abundance of marshes.

The role of continental glaciation in forming the Great Lakes has already been discussed. Moraines, outwash plains, eskers, drumlins, and glacial erratics supply irrefutable evidence of the presence of a vast ice sheet and the colder climate which accompanied it. The geological period marked by advances of continental ice sheets over large portions of the Northern Hemisphere (the Pleistocene) began sometime around 1 or 2 million years ago. It was characterized globally by temperatures which were generally colder than those which exist today and by a number of episodes (possibly as many as eight) during which ice advances occurred. These ice advances seemed to begin and end with remarkable rapidity. At their termination, brief interglacial periods with warm temperatures persisted for a short while, followed by a decline of temperatures and return of the ice.

The last advance of continental ice sheets over North America is known as the Wisconsin stage. It began about 20,000 or 25,000 years ago and was responsible for the formation of the Great Lakes (see chapter 2). About 11,000 years B.P. (before present), the climate of the region changed from the cold, snowy conditions which accompanied the retreat of the glacier to a warmer, drier, less snowy climate of an interglacial period. This last interglacial, which includes the present, is known as the Holocene. Mean global temperatures during the Holocene have been about 7°F to 11°F (4°C to 6°C) higher than at the height of glaciation, although in some areas the difference is larger and in other areas it is less (Bryson 1974).

It is important to place the climate of the Holocene in perspective. First, we must consider that the Holocene has been warmer than the majority of the past million years. Interglacials have broken the cold pattern dominant during the Pleistocene for only brief intervals. Second, present evidence suggests that the interglacials lasted from 7,000 to 10,000 years. Thus the probability of the present interglacial ending soon is higher than we might like, at least if past climatic behavior is any indication. Third, while the general pattern of global climate during the past 10,000 or 12,000 years has been relatively warm, significant fluctuations of temperatures and weather patterns have occurred within the Holocene—and some of them appear particularly well marked in the Great Lakes area.

Climatic changes during the Holocene

Since the changes which occurred during the Holocene were of an order of magnitude smaller than those during the Pleistocene, they are a bit more difficult to detect and evaluate. In the Great Lakes region, instrumental records are available only from the past 150 years, and historical data (inferences regarding past weather from written records of freezes and thaws of rivers, harvest quality, unusual weather events, etc.) are not available for a much longer period. The climatic changes of the Holocene, up to the time when historical or instrumental data became available, must be examined through the use of *proxy data;* that is, data from natural systems which, when carefully analyzed, may provide some information about the climate of the past. These techniques may include the analysis of tree rings (dendrochronology), oxygen isotopes in ocean sediments, fossils, ice cores, and varved (seasonally layered) lake sediments.

Pollen analysis (palynology) utilizes proxy data, and the use of palynological techniques has enabled paleoclimatologists to learn something about the changes of climate which have occurred in the upper Midwest since the retreat of the Wisconsin ice sheet. Pollen analysis can be used to reconstruct past climates when layered sequences are obtained. The frequencies of various pollen types found in the layers can tell much about the vegetation pattern which existed at the time the layer was deposited. Changes of individual pollen-type frequencies in different layers of the sequence indicate changes in the complex

Winter of 1976–77 in the Great Lakes. Sandy Creek, New York, 21 January 1977.

of vegetation from time to time. The basic premise, of course, is that the vegetation complex is a very good indicator of the type of climate which exists.

Some caution must be used in interpreting pollen frequencies. For example, a future scientist utilizing this technique to learn the present climate of the Great Lakes might be misled by the reconstruction of the vegetation cover which palynology might provide. The vegetation type which currently exists has been highly altered by man and may be only partially indicative of climatic factors. In addition, plant associations may be in the process of reaching an equilibrium with climate, and therefore the pollen analysis indicates only a stage within a succession toward a final equilibrium situation. In spite of these handicaps, if one proceeds with caution, much can be learned about past climates from an examination of pollen sequences.

Analyzing fossil pollen sequences from the beds of three lakes in the upper Midwest, scientists from the University of Wisconsin and the University of Michigan have been able to piece together some of the climatic events which have occurred since the retreat of the ice sheet. Pollen analysis indicated that a change from glacial conditions to the Holocene occurred about 11,300 B.P. (Webb and Bryson 1972). During the late glacial period, air masses from the northwest dominated most of the year with cold, snowy conditions. As the transition to the Holocene occurred, westerly air masses from the Pacific became more frequent and July temperatures increased.

From 9500 to 4700 B.P., the frequency of westerly air masses increased. The westerly air masses brought drier conditions eastward toward the Great Lakes area. Today a region of prairie vegetation extends toward the Great Lakes like a wedge from the base of the Rockies. Maximum westerly flow, according to the results of the fossil pollen analysis, occurred about 7200 B.P. This was a rather warm period globally, probably warmer than at any time since the retreat of the Pleistocene ice sheets. The period has been called the *climatic optimum** and was characterized by a high frequency of westerly jet stream situations. In the upper Midwest, drier conditions existed because the increased flow of westerlies prevented incursions of moisture-laden air from the south, and cyclone development was also weaker.

The climatic change which marked the transition between the Pleistocene and the climatic optimum was larger than any of the subsequent changes occurring within the Holocene. The general tendency has been toward cooler conditions during the most recent 2,000 years. Paleoclimatologists are able to make use of another concept called *teleconnection* to reconstruct the climatic patterns in the Great Lakes during the recent millenium.

The concept of teleconnection assumes that climatic change, whatever its basic cause, is manifested through changed air circulation patterns. These may involve contractions and expansions of the circumpolar vortex (westerlies with jet stream) and changes from low to high zonal index. With high zonal index, the westerlies flow mostly from west to east. With low zonal index, the westerlies loop poleward and equatorward in waves. During the warm-up of the climatic optimum, the upper air westerlies assumed high-zonal-index flow characteristics. At later periods in the Holocene, the westerlies aloft were best described by a lower zonal index (figure 71). With low-zonal-index situations, regions located within troughs tend to cool, while regions located beneath ridges tend to warm.

Since the ridge-trough complex in the northern hemisphere is anchored by the mountains of western North America, some downstream areas will more often be affected by troughs during low-zonal-index situations, others by ridges. For example, eastern North America and western Europe tend to be located within troughs when the atmosphere shows a long-term ten-

---

*Recently there have been some objections by climatologists to the use of this term. Optimum for what? they ask.

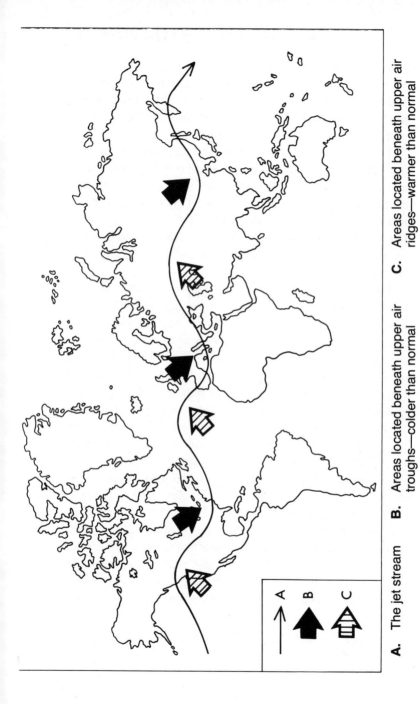

A. The jet stream

B. Areas located beneath upper air troughs—colder than normal

C. Areas located beneath upper air ridges—warmer than normal

Figure 71 **Teleconnections and the Jet Stream**

dency toward a low-zonal-index circulation pattern; consequently, it is likely that a teleconnection exists; that is, that the climatic changes, in response to the circulation change, occur in the same direction. As western Europe experiences a cooling trend, so does eastern North America.

Although the teleconnection concept does not always give an accurate estimation of climatic conditions, in many cases rational inferences can be made. For a reconstruction of past climates in the Great Lakes area, it can become useful that instrumental records and written historical materials are available for a considerably longer period in Europe than in North America. This means that if we are able to reconstruct the climate of western Europe with some degree of accuracy, we can also make certain inferences about the climate of eastern North America.

In Europe, the general climatic features for the past 1,000 years have been reconstructed with considerable accuracy. Professor H. H. Lamb and his climatological group at the University of East Anglia, United Kingdom, have painstakingly examined manuscript sources, historical accounts, various proxy data sources, and, of course, instrumental data, in reconstructing the climatic patterns of Europe and the North Atlantic for the past millenium. From the findings of this group, we are able to infer certain conclusions about the climate of the Great Lakes through the teleconnection concept. Instrumental data of the past 100 years or so, of course, are available for the Great Lakes area.

In Europe, the period A.D. 900–1300 appears to have been warm—not quite comparable in warmth to the climatic optimum, but nevertheless a period of mildness sufficient to warrant being called the *secondary climatic optimum*. During this warm period, the Gulf Stream shifted northward in the North Atlantic, the Arctic pack ice retreated into higher latitudes, mountain glaciers retreated and many disappeared, and the westerlies aloft assumed a more zonal alignment, thus transporting mild Atlantic air far into the interior of Europe. It was during this warm period that the Vikings were able to colonize Iceland and Greenland and that some agriculture could be supported in these areas. Grapes grew in Great Britain, where the climate is too cool today, and reports of grapes growing in Labrador (whereby the name Vineland was derived) may well have been based on fact.

After A.D. 1300, all indications point toward a progressively colder climate in Europe. The westerlies returned to a low-zonal-index configuration, with troughs over the western portion of the Atlantic and over Europe. The coast of Iceland became encased with pack ice for many months during the year, and the climate of Greenland became colder. Glaciers in the mountains began to advance down their slopes and new glaciers formed. The Viking colonies perished as agriculture was no longer feasible. Rivers in Europe which normally remained ice-free the year round froze over with unparalled frequency as winters became colder and more snowy.

Teleconnections between Europe and eastern North America suggest that the latter area was also cold. The equatorward dip of the upper air westerlies over eastern North America increased in amplitude, and cold air masses spawned in Canada entered the United States with increasing frequency. The Great Lakes area, immediately downstream from the forcing action of the Rocky Mountains, was positioned at a critical point. Small changes in circulation could become amplified because of the sensitive location of the Great Lakes, and an unambiguous response in the form of climatic change was the likely result.

This cold period was known as the *little ice age*. The little ice age continued, with smaller-scale fluctuations, into the nineteenth century. The eastern seaboard of North America was colonized during the little ice age, and the early colonists suffered the additional hardship of arriving during a period which now appears to have been the coldest of the past several millenia. Narrative and diary descriptions of heavy snows and severe frosts in areas where today these features occur infrequently allude to the more severe climate which the colonists faced.

Climatic change during the past 100 years in the Great Lakes

Some instrumental data have been available in the Great Lakes area since the early 1800s and from the eastern United States, at some sites, since the 1700s. Instrumental records allow climatologists to reconstruct the climatic patterns in eastern North America and the Great Lakes area for the past 100 years or so. But even with instrumental data available, the job is not an easy one. Instrumental data must be interpreted with a great deal of caution, for many problems may make the records invalid for studying climatic change.

For example, most long-term instrumental records come from cities, and the growth and outward spread of the cities themselves can introduce apparent trends in the data which are not the result of climatic change. We will discuss the ways by which cities modify climates in a later section. Changes in siting of instruments, observers, changes in observer technique and the natural environment—all can introduce extraneous factors into the interpretation of long-term data records.

Nevertheless, with proper care, instrumental records can be and have been used to study the changing climate. In the Great Lakes area, studies have shown that the region has responded quite strongly to Northern Hemispheric trends in temperature and circulation during the past 100 years. The Northern Hemisphere as a whole has witnessed increasing temperatures during the latter decades of the nineteenth century, continuing until about 1940 or 1950. While the little ice age was characterized by a low-zonal-index pattern in the upper westerlies, with marked equatorward loops and poleward meanders, the warming trend was accompanied by a change to high zonal index with stronger westerly flow.

Here again the Great Lakes, in a sensitive location, underwent a change of significant dimension. While during the early 1800s the climate of the region was cooler and wetter in the summer and colder and snowier in the winter, an upward trend in summer and winter temperatures occurred during the latter decades of the nineteenth century. Precipitation also decreased, related to the more prevailing westerly flow and the northward displacement of the polar front in summer. In some ways, conditions suggested a return to the climatic regime which occurred during the early Holocene at the time of the climatic optimum about 7200 B.P.—or during the secondary climatic optimum 1,000 years ago.

The rising summer temperatures of the late 1800s and early 1900s seemed to culminate in the Great Lakes region during the 1930s, while rising winter temperatures culminated in the 1940s and 1950s. After 1950, hemispheric low-zonal-index situations occurred with increasing frequency. The change from the westerly pattern of the early 1900s to the low-zonal-index pattern of the latter decades occurred with a suddenness that prompted Lamb (1966) to comment that while such changes had occasionally occurred in the past, they were rare.

The significance of the warm-up during the early 1900s resides in the following considerations:

First, our records of climatic normals are taken from 30-year periods. The 30-year normal period 1931–60 embraced the peak of the twentieth-century warm-up. Recent evidence has suggested that temperatures in the 1930s, at least in some parts of Europe, were warmer than during the past 1,000 years. Assuming that the Great Lakes region responds in phase with western Europe and with hemispheric changes (but more vigorously), we may draw a tentative conclusion: For the Great Lakes area, the 1930s and 1940s were actually highly unusual climatic periods—possibly warmer than for the majority of time during the past 1,000 years. So the normals extracted from the

1931–60 period were really abnormal and not representative of the climate over a longer period.

Second, the burgeoning population, industrial growth, and, especially, the recreational and agricultural development within the Great Lakes area has occurred within a climatic regime which now appears to be abnormal.

Third, definite signs exist that the climate is returning, or perhaps has returned, to a pattern reminiscent of the early 1800s or to the more normal pattern which has characterized the past 1,000 years. This statement bears closer examination.

Climatic changes during recent decades

In line with the hemispheric cool-off, the Great Lakes area has also experienced a cooling trend since the 1930s and 1940s. But this tendency toward cooler weather has not been equally pronounced in all sections of the Great Lakes, nor have all seasons of the year become cooler. The southern portions of the Great Lakes seem to have shown the most marked fluctuation during the past several decades. This regionalization of change can be of major importance, because it is the southern Great Lakes area which contains the majority of the population, and this is also the region of greatest agricultural significance.

The recent change of climate in the southern Great Lakes area seems most pronounced during the middle of winter and at the height of summer. January and July temperatures have shown significant declines. The transition months, spring and fall, have not shared in this cooling tendency; in fact, some spring and fall months have actually become warmer. Accompanying the drop-off of summer temperatures have come increased cloudiness, decreased amounts of sunshine, and increased July precipitation (figures 72–74). The downward trend in summer temperatures has occurred primarily as a result of decreases in the *daily mean maxima*. What this means is that afternoon temperatures are not quite as hot as during the height of the warm-up in the 1930s. With fewer clear days and more

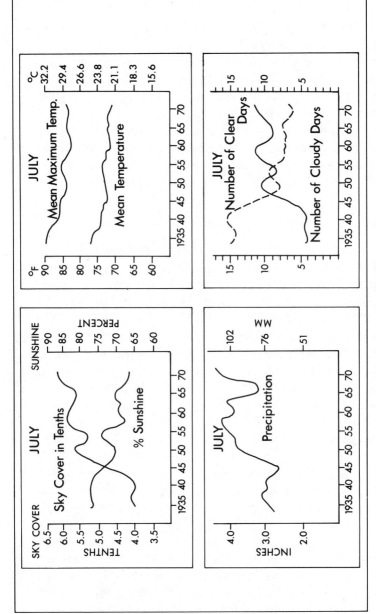

Figure 72   Trends of July Climate for the Southern Great Lakes–Eastern Corn Belt Areas

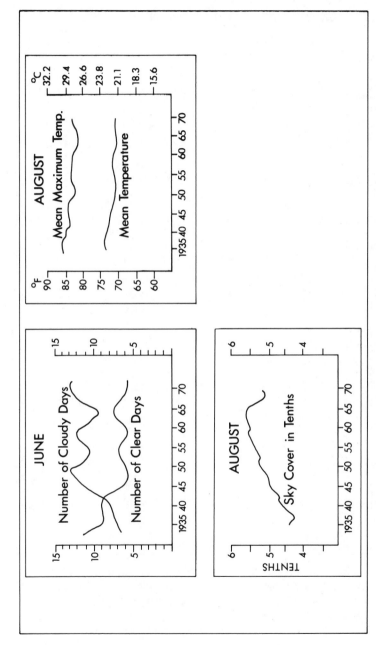

Figure 73 Trends of Summer Climate for the Southern Great Lakes–Eastern Corn Belt Areas

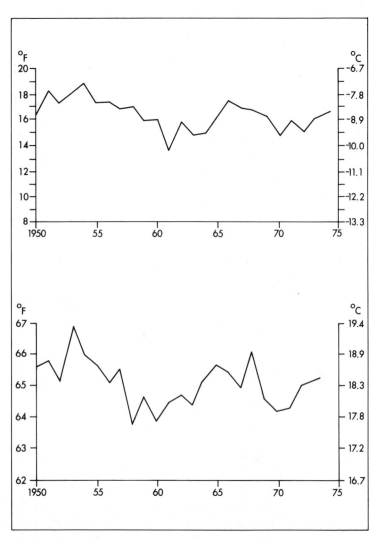

Figure 74 Five-Year Moving Means of January *(above)* **and July** *(below)* **Temperatures for Chatham, Michigan, in the Upper Great Lakes Region. Note that the downward trend of temperatures occurring in the southern Lakes region is not readily apparent in the northern region.**

cloudy days, strong indications exist that *if* this tendency continues, the summer climate will be returning to that which existed during the early and middle 1800s.

If this happens, both major and minor adjustments may have to be made. Predictions of future agricultural yields and the selection of cropping practices must be done with the current climatic fluctuation in mind. It is possible that, for awhile, cooler summers might be beneficial to agriculture for the reason that the reduced evaporation and increased precipitation may increase moisture availability. Grain crops should do well, as they have for a succession of years up to the early 1970s. But if the trend continues unabated, more unfavorable effects are sure to follow. A shortening of the growing season has already been noticed at some stations in the Midwest. And the decrease of solar energy coincident with lessened summer sunshine must also be a consideration in the future estimation of crop yield.

The effects of cooler summer temperatures and less sunshine also must affect the recreation and tourism industries which are so important in the southern Great Lakes area. And future projections of energy demand must be made within the framework of changing climate. Cooler summers mean less demand for air conditioning and possibly less vacation driving. However, colder winters mean larger demands for heating fuel and possibly more driving to ski resorts.

No one can predict what the future will bring in the way of climate for residents of the Great Lakes area. Some indications exist that the summer and winter cooling pattern for the last several decades may have leveled off and perhaps even reversed itself. Some scientists feel that global warming is already occurring and that the major climatic problem during the next few decades will be drought and excessive warmth. Weather patterns during the early and middle 1970s appeared rather different from those of the 1960s.* Perhaps this is only a

*However, the winter of 1976–77 in the Great Lakes area had all the earmarks of the weather patterns which dominated the 1960s.

temporary departure from a long-lasting trend toward cooler weather for the area. Or perhaps it marks a reversal of the trend of the past several decades. It is simply too early to tell. It does seem clear, however, that the summer weather of the 1930s and 1940s was highly unusual and perhaps represented an abnormality unmatched during the past 1,000 years. The cooler, cloudier summers of the 1960s and 1970s seem to be a step closer to normality, if one is bold enough to employ such a term when describing climate.

PART FOUR

Humans' Role in Altering the Weather of the Great Lakes

15: Humans' Inadvertent Modifications of Weather and Climate

WE HAVE KNOWN FOR SOME TIME THAT WHEN PEOPLE congregate in cities, as in the Great Lakes area, subtle weather changes occur. Studies of European cities made in the nineteenth century showed them warmer and less sunny than the surrounding rural areas. Such weather changes are inadvertently caused. They result accidentally from the activities of man and stand in contrast to the deliberate modifications which man attempts through rain making, cloud seeding, and other planned activities. Although inadvertent modifications result without a conscious attempt by man to change the weather, the changes wrought are probably more substantial and widespread than those resulting from deliberate efforts.

Although inadvertent weather modification has become a problem of serious dimensions only in recent decades as world population has rapidly expanded and cities have stretched their boundaries outward toward rural areas, we can find some prime examples in the past. Professors Bryson and Baerreis (1967) of the University of Wisconsin have investigated the Harappan culture which occupied what is now the Rajputana desert area of northwest India about 2,000 B.C. The Harappan civilization was an advanced, prosperous culture when inexplicable de-

244

terioration began and the Harappans succumbed to invasions by alien tribes. Bryson and Baerreis presented a new theory to account for the demise of the Harappans. The Harappans, they maintained, had caused their own downfall by inadvertently altering their climate. Available evidence suggests that what is now a desert area was a grassland at the time of the Harappan ascendency. It was a region with considerably more rainfall than at present and was capable of supporting some agriculture. But by plowing the land and removing the grass cover, the Harappans introduced dust into the atmosphere. This, in turn, altered the vertical temperature distribution of the atmosphere to such an extent that precipitation was suppressed. With less precipitation, the area reverted to a desert, agriculture failed, and the Harappan culture disappeared.

The La Porte Anomaly

More recently, a celebrated example involving a city within the Great Lakes region alerted both scientists and laymen to the fact than man can alter the weather without even trying. Probably no single event has aroused as much interest or controversy regarding inadvertent climatic modification as the *La Porte anomaly*.

La Porte, a medium-sized city in northwestern Indiana, presented a rather strange precipitation record (figure 75). The climatic record from 1925 until about 1960 had shown a 30% to 40% increase in precipitation. None of the surrounding stations shared this increase. The graph of precipitation over the years appeared to be in phase with the production curve for steel in the South Chicago–Gary iron and steel industrial complex 30 miles to the west. Industrial areas supply heat to the atmosphere, and ice-forming nuclei which play a role in the precipitation process are also added by iron and steel making. Thus there was a suggestion of causal connection between the industrial

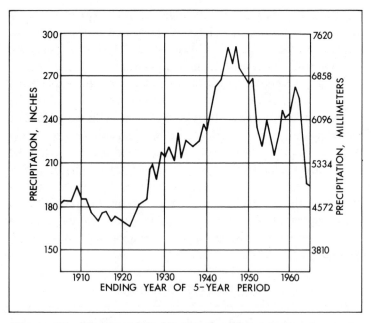

Figure 75   Five-Year Moving Totals of Precipitation Amounts at La Porte, Indiana

activities west of La Porte and the increase of precipitation at La Porte.

In an attempt to determine whether the increase at La Porte was real (and thus probably an example of inadvertent modification) or whether the apparent increase was fictional (the result of observer bias or siting changes of the rain gauge), Stanley Changnon, Jr., of the Illinois State Water Survey made a careful, unbiased examination of the La Porte record. Publishing his results in the *Bulletin of the American Meteorological Society,* he concluded that the increase was a real one. He also found that La Porte had more thunderstorms and hail than surrounding stations (Changnon 1968*a*).

Changnon asserted that the La Porte situation was valid proof that cities could alter the rainfall patterns over and downwind of them. In addition to alerting the scientific community of this fact, news of the La Porte anomaly also caught the public eye. *Saturday Review, Scientific American, Industrial Research, Newsweek,* and *News Focus* all carried stories of the change of climate which industry was suspected of causing at La Porte.

The news of the La Porte anomaly did not go unchallenged in the scientific world. A study examining the climatic records in the vicinity of the Port Kembla, Australia, steel plant concluded that the rainfall was not influenced by the steel works. Other scientists refuted the existence of a La Porte anomaly, attributing the apparent rainfall increase to observer error. A long series of exchanges appeared in the scientific journals, and when the controversy finally died, much evidence had been presented to suggest that the anomaly was fact and not fiction—but there was no irrefutable proof.

More important than the controversies, however, was the interest aroused over the La Porte data. Scientists and laymen alike suddenly became aware of the potential for alteration of the atmosphere by man. Several research proposals were funded, including a 2-year grant to the Illinois State Water Survey for a study of precipitation patterns around eight major American cities, and the joint 5-year METROMEX project to study precipitation patterns in the St. Louis area. The results of these studies have added much to our knowledge of how large urban areas can affect weather and climate.

So it was that a medium-sized city in the Great Lakes region focused attention on man's capacity to alter the delicate atmospheric balance. Of particular interest are the unique locational aspects of La Porte, and the presence of one of the Great Lakes (Lake Michigan) not more than 15 miles (24 kilometers) away. Much evidence suggests that the presence of the lake has played a strong role in creating the anomaly. The cold water of the lake and the lake-breeze front, along with the industrial effects from upwind, are likely factors in increasing precipitation in the La Porte area. Much evidence also indicates that the presence of the Lakes may be a contributing factor to other examples of inadvertent weather modification. We will examine some of this evidence by looking at some of the ways by which man has changed the climate in the Great Lakes region.

Climatic Alterations by Cities

Characteristics of the surfaces of cities, particularly large ones, are such that the weather and climate in the immediate area and downwind may be changed, in some cases significantly. As the Great Lakes area contains a number of very large cities, it is important that we look at some of the ways by which the presence of a city alters the climate. Additionally, we will see that the presence of the Great Lakes interacts in various ways with the city-caused modifications.

City surfaces versus rural surfaces

The surface of the city differs from that of the rural countryside (figure 76). While the rural areas are likely to have surfaces composed of natural materials such as soil or sod with many trees and shrubs, city materials are mostly artificial, consisting of concrete, asphalt, roofing materials, and so on.

In what ways might this difference in surface materials affect weather and climate? One way is rather obvious. Precipitation in the city is received on an impermeable surface. There are no soil pore spaces to hold water as in the countryside. Thus the water runs off into sewers and storm drains. This means that less evaporation will occur near the surface. Evaporation supplies moisture to the air and is also a means of cooling, since energy is used in the evaporation process. Thus the atmosphere over the city tends to have a slightly lower relative humidity, particularly in the afternoon and an effective cooling

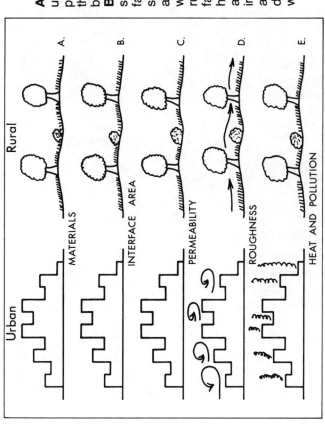

A. The greater thermal conductivity of urban surfaces (e.g., concrete, asphalt) causes them to heat less rapidly than rural surfaces (e.g., grass, trees) but to store large amounts of heat. **B.** The vertical components of urban surfaces create a more extensive interface with the atmosphere. **C.** Urban surfaces consist of impervious materials. Rainfall runs off such surfaces, while infiltrating the more pervious rural surfaces. **D.** Rougher urban surfaces induce turbulence and reduce horizontal wind velocities. **E.** Large amounts of artificial heat are released into the city atmosphere from domestic and industrial heating units, air conditioners, etc. Pollution is also greater within urban areas.

Figure 76 Urban versus Rural Surfaces

Plumes from the South Chicago–Gary area initiating cloud development over Lake Michigan. An example of inadvertent modification. ERTS I satellite photograph 1003 CST, 24 November 1972.

Smoke and cloud at a lake-breeze front—southern end of Lake Michigan.

Cohesive plume heads east over Lake Michigan during conduction inversion.

mechanism which is present in the country is absent within the city.

There is another way in which the difference in surface materials between the city and the country is important. Remember the concept of thermal conductivity which explained the contrast in heating rates exhibited by land and water bodies? The artificial materials of the cities have higher thermal conductivities than those of the country. This means that heat is diffused downward more rapidly within the concrete and asphalt surfaces of the city than within the natural surfaces of the country. While the materials of the country may warm a bit more rapidly than those of the city, the city materials have a much larger capacity to *store* heat. This heat may then be released into the atmosphere, thus raising its temperature.

The impervious nature of the city surface and its larger thermal conductivity combine to cause the city to be warmer at times than the country. Consider also that the surface of the city tends to be more extensive than that of the country (figure 76); that is, the city's very irregular surface consisting of buildings presents more surface area as an interface to the atmosphere. With a large thermal conductivity combined with a greater surface area, the city absorbs and stores more heat than the country.

The rougher interface of the city means that winds encounter greater friction in flowing over the city. The added friction, in turn, means that overall wind velocity will be reduced (although it may be windier at street level) and also that added turbulence will occur. Turbulence in the form of updrafts causes a lifting motion of the air flowing over the city surface. Lifting motion and added heat, remember, are conducive to cloud formation.

The city interface also injects heat into the atmosphere due to the loss of heat from furnaces, heating units, air conditioning, exhausts from automobiles, heat loss from engines, and many other sources. Along with this artificially produced heat supplied to the city atmosphere is a variety of pollutants largely absent in the country.

The heat island

Probably the most conspicuous climatic feature which stems from these differences is the *heat island*. This phenomenon occurs not only with large cities but also with medium and small cities. It has even been detected when the temperatures around shopping centers have been monitored.

The heat island, though not always present, is most prominent at night and under fair weather conditions with light winds. At these times, some rather large differences in temperature have been recorded between the center of the city and the suburbs. Great Lakes cities are no exception.

While the heat island is the most prominent modification induced by the unique characteristics of city surfaces, other modifications may stem from the heat island itself. The presence of a warmer zone near the earth's surface leads to increased convection and uplift. Consequently, the frequency of afternoon clouds over and downwind of the city may be increased. When the overall winds are very light, the heat island may instigate a wind field much like that of the lake breeze. With the lake breeze, the difference in temperature between the lake and the land causes a breeze to blow toward the land during the day. In the city, when the heat island exists, local pressure differences may cause winds near the surface to flow toward the center of the city. There they rise, then return aloft toward the suburbs (figure 77). This heat island-induced circulation may be effective in dispersing pollutants vertically into the atmosphere but may also add to the uplift of air which occurs over the city, thus increasing the amount of cloudiness.

The pollutants emitted by the industries and automobiles of the city may cause an abundance of nuclei for the condensation of cloud droplets, thus adding to the possibility of increased

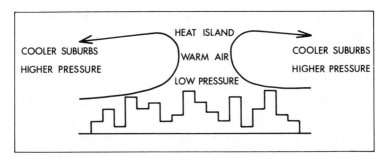

Figure 77   City-Induced Wind Circulation

cloudiness in the city. Additionally, the pollution of the city decreases visibility and reduces the amount of solar radiation (particularly of the ultraviolet range) reaching the city. This reduction is particularly marked during the winter months when the sun is at a low angle.

The combination of added heat in the city, altered wind fields, and pollution can cause increased fog, increases in thunderstorms, hail, and rainfall, as well as distinct temperature contrasts. These are climatic alterations which seem to be common to most large cities.

Some modifications by Great Lakes cities

In the Great Lakes area, studies have indicated some significant modifications of weather and climate around large cities. The influence of the Chicago urban area has already been discussed. La Porte, Indiana, downwind from Chicago received 17% more rainfall than Chicago in the 1949–68 period. More hail and thunderstorms also seem to occur in northwestern Indiana. In the Chicago area itself, the heat island is apparent on many occasions during the year. Studies of thunderstorm occurrences (Huff and Changnon 1972) have shown that the urban airport (Midway) experienced more thunderstorms than the rural airport (O'Hare International). The presence of Lake Michigan just to the east introduces lake effects which may enhance or suppress the city effects depending on the time of year.

Urban enhancement of precipitation has been noted in an area east of Cleveland, Ohio. As in Chicago, the enhancement is most pronounced on days when rainfall is moderate to heavy. At Cleveland also, the lake effects tend to mask the city's role in altering the climate.

A study of the Detroit metropolitan area (Smith 1966) has shown that the heat island has different diurnal and seasonal locations because of the influence of Lake St. Clair to the east and north. When the temperature of the urban area was warmer than the lake, the lake acted as a cooling agent; and when the temperature of the urban area was less than the lake, the lake created a warming effect. The symmetry and location of the heat island was thus influenced by the surface temperature of the lake.

Detroit City Airport, located near the center of the metropolitan area, showed warmer temperatures on most days than did Detroit Metropolitan Airport in the western suburbs (figure 78). In addition, the influences of the lake and the city combined to retard the occurrence of the first frost during the fall; and, in the spring, to cause earlier occurrence of the last frost (figure 79).

Increases in rainfall also occurred downwind from Detroit when convective systems were in the area, owing to the additional uplift caused by heat from the city plus westward penetration of the lake-breeze front from Lake St. Clair (Eichenlaub and Bacon 1974).

Large numbers of ice-forming nuclei have been detected over and downwind of steel production areas near Buffalo, New York, and Hamilton and Toronto, Ontario; and initiation of snow shower activity has been observed from this pollution. The large numbers of automobiles in the Great Lakes area have

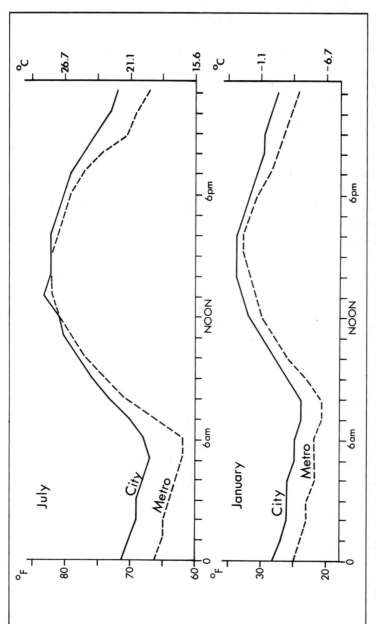

Figure 78   Average Hourly Temperatures at Detroit City and Metropolitan (suburban) Airports in January and July 1964

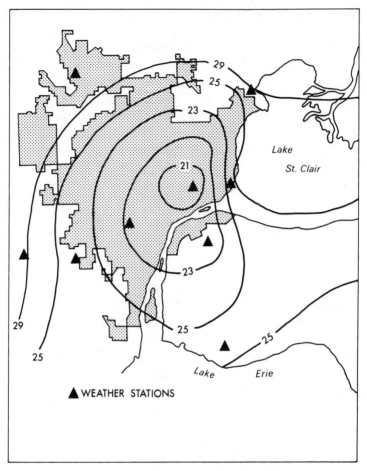

**Figure 79   Mean Date in April 1959–65 of Last
Frost Day of Spring**

also been linked to the production of ice-forming nuclei over
and downwind of the densely populated lower Lakes area,
formations due to the lead in automobile exhaust which
combines with iodine in the atmosphere. The lead-iodine
compound thus formed serves as an excellent nuclei for the
formation of ice crystals.

Certainly the evidence is ample that Great Lakes cities have the capability of altering the weather and climate around them. The particular problem in identifying and quantifying these alterations is the interaction of the Great Lakes, which create their own set of effects. At times the lake effects combine with the urban effects to produce the observed alteration. At other times, the lake effects operate in opposition to the city effects. Cities and lake effects simply add another link to the chain of complexity which is Great Lakes weather.

Nuclear Plants along the Shores of the Great Lakes and Possible Inadvertent Modifications

With its large population and strong industrial and agricultural base, the demand for power in the Great Lakes area will continue to grow. Construction of nuclear plants has been seen as a possible partial solution to the problem of supplying ever-increasing energy demands. Nuclear power plants need large amounts of water for cooling. With their abundant supply of fresh water, the shores of the Great Lakes have become an attractive region for the construction of nuclear plants. A number of them have been constructed, and more are in the planning stages.

The meteorological problem posed by the nuclear plants concerns the discharge of the waste water used for cooling purposes. During the process of using it as a coolant, the water acquires a considerably higher temperature. It must then be disposed of in a manner which has the least environmental impact. Although many facets of the environment might be altered in the disposal process, our concern is with the meteorological and climatological problems.

There are essentially two ways by which the water may be disposed of. One way is simply to send the water into a *once-through process;* that is, take it from the lake, send it through the condensers, then return it to the lake. Another way is to

Cooling tower (not in operation) of Toledo Edison's Davis-Besse nuclear plant.

utilize cooling towers. The purpose of the cooling tower is to extract the heat from the water and dissipate it into the atmosphere. Cooling towers are usually (though not always) very tall structures (about 250 feet [76 meters] in height) where the heat is extracted either by evaporation or by conduction and convection. A cooling tower in which evaporation is the main means of cooling is known as a *wet type;* where conduction and convection are the processes used, the tower is known as a *dry type*. With wet-type cooling towers, both heat and moisture are added to the atmosphere. With the dry-type cooling tower, heat, but little moisture, is added. Each method of· waste water management has a potential for altering the weather and climate, and the lake effects interact with these modifications as they do with the urban effects.

Possible modifications resulting from the once-through process

Let us examine first the possible changes which might occur with return of the heated water directly to the lake. Strong protests came from environmental groups when this method of water disposal was originally proposed. The environmental groups foresaw widespread alteration of the marine environment because of increased water temperatures. They decried the possible effects on fish and plant life in the Lakes, but the possible atmospheric effects were generally overlooked. In reply to the environmentalists, the power companies, making their own assessments, insisted that the heat would be dispersed into the lake and dissipated into the atmosphere so quickly that few if any changes in the marine environment would result. Possible atmospheric effects of the heat dispersion were largely ignored.

Based on theory and on a recognition of how the Lakes affect climate, we may be able to make a few basic predictions of how the atmosphere might be altered by the once-through method. During the unstable season, when the lake is warmer than the land, all lake effects would tend to be amplified. The amount of heat supplied to the atmosphere from the lake would be increased, as would the transfer of moisture, with the addition of the warm-water discharge. Cloud-forming processes would be augmented and the possibility of increased lake-effect snow must be considered.

Other environmental changes might also occur. The discharge of warm water into the lake might also prevent the formation of protective shore ice. The shoreline, if it remained unprotected and exposed to the pounding of waves and currents throughout the winter, would be subject to increased storm erosion. Also, during the late winter, the decrease in the extent of open lake water, which normally occurs because of the buildup of shore ice, might be forestalled.

During the stable season, the presence of warm water might lessen the temperature contrast between the lake and the land, thus decreasing the occurrence of the lake breezes. Convective cloud formation and convective rainfall, which are normally suppressed by the lake, might become more frequent, and nighttime thunderstorm activity and shoreline hail storms might occur more often.

These are modifications which we can foresee when theory is invoked. The question is, to what degree would these events actually occur? We cannot answer this question— meteorologists simply do not know. Best-guess estimates state that with one or two plants utilizing the once-through method, the changes of climate would be very small, almost insignificant. The real problems might arise with the establishment of a large number of plants within a limited area, such as around the perimeter of southern Lake Michigan. It is impossible to predict the threshold when inconsequential changes in climate might become consequential. Also, who is to say what is inconsequential or not? Further research aimed at recognition of this point is necessary.

Modifications which may result
from cooling towers

The cooling tower alternative would appear to avoid many of the environmental changes which worry the environmentalists. Discharge water would be returned to the lake at approximately the temperature at which it was originally taken. Biotic life within the lake would be unaffected. But how about the atmosphere? With cooling towers, transfer of heat and moisture (with wet-type coolers) would occur directly into the atmosphere without the benefit of dispersal which might be provided by currents in the lake. Heat and moisture would then be supplied to the air from a more concentrated source. In theory, the addition of heat and moisture from cooling towers might increase the frequency of fogs, clouds, icing, and precipitation immediately downwind. With the frequent occurrence of conduction inversions in the summer over the Great Lakes, these effects could be amplified due to the trapping effect of the inversion.

Little is known about the quantitative dimensions of these theoretical modifications. A pilot study of the Zion Nuclear Plant along the shores of Lake Michigan in northeastern Illinois (Huff et al. 1971) has given some information regarding the significance of these changes. Zion is a 2,200-megawatt plant with three wet-type coolers over 250 feet (76 meters) in height. Thus both heat and moisture are emitted from the towers into the atmosphere. Heat emission from the Zion coolers was estimated as 16% of that for the city of St. Louis and 5% of that for Chicago. Thus the cooling towers act as small cities, supplying heat which may play a role in convective uplift.

The Zion study predicted that the cooling towers would cause an increase in fog but only for 1 or 2 miles (1.6–3.2 kilometers) from the lakeshore. The total annual snowfall might also increase 1 to 2 inches (2.5–5 centimeters) along the immediate lakeshore. Under favorable conditions, thunderstorms might be triggered, but the total increase in precipitation caused by the cooling towers would amount to only a fraction of 1%.

Studies conducted by University of Michigan scientists (Ryznar and Weber 1976) indicate that downwash from cooling towers of the Palisades Nuclear Plant near South Haven might double the number of hours of heavy fog occurring in the immediate vicinity of the plant. Effects downwind would be slight, however.

Small changes? Yes. Are they insignificant? Probably, but who is to judge? These are changes associated with single nuclear plants and their batteries of cooling towers. What would be the effect of twenty nuclear plants?

We should keep several points in mind before we leave this discussion of cooling towers. First, the Zion study concluded that the meteorological effects would probably be smaller if the once-through process was used, with the warmed water first dissipated into Lake Michigan, then from the lake into the air. The intermediate step of diffusion into Lake Michigan would remove the concentration of heat and moisture which the cooling tower provides. A second point to remember is that the presence of the Lakes introduces some very special meteorological effects (such as the summertime conduction inversion) which must be considered whenever we attempt to appraise and predict the inadvertent modifications of weather and climate.

The Great Lakes and Atmospheric Pollution

With its many cities and industries, the Great Lakes area (especially the southern portions) has not escaped the problems of air pollution. While the region's frequent fronts, cyclone passage and air mass exchange would appear to favor the quick vertical and horizontal dispersal of atmospheric pollutants, certain factors may tend to magnify the problems associated with air pollution.

Some of these factors are nonmeteorological and concern the tendency of people, cities, and industrial areas to congregate near the shores of the Lakes. For heavy industries, location near the Lakes may be desirable because of the cheap transport of bulk materials which the Lakes afford and the availability of fresh water for industrial uses. With continuing growth in the lower Great Lakes area, the problems of clustering along the lakeshore may be expected to grow.

Other factors compounding the air pollution problems in the Great Lakes area are primarily meteorological and are linked to the special weather effects which the Lakes can produce. Professors Cole and Lyons (1972) of the University of Wisconsin–Milwaukee have outlined and explained the relationships between lake effects and pollution. Noting the role of the lake during the stable season, they have examined the effectiveness of pollution dispersal into the atmosphere during conditions when a stable onshore flow of air from the lake may occur.

In lake-breeze conditions, these researchers have photographed plumes directed onshore by the stable, upper portions of the lake breeze. Under sunny conditions, a mixing layer develops as the cooler lake breeze encounters the warmed land surface. The mixing layer deepens as the lake breeze moves farther inland. The mixing level may build up to the height of the plume, in which case the plume is brought downward, with a trapping effect (figure 80). Near the lake-breeze convergence zone, walls of smoke have been detected; while aloft, above the lake-breeze advance, pollutants are transported back toward the lake. As a result of the lake breeze, higher concentrations of pollutants are found in some shore areas as vertical dispersal is prevented. In situations where lake breezes are absent but a flow of cool, stable air from the lake occurs because of the synoptic scale wind field, the effects can be the same. With cloudy skies, the mixing level near the surface may remain very shallow, effectively trapping pollutants close to the ground.

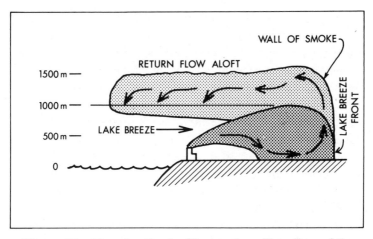

Figure 80   Trapping Smoke Plumes along Shorelines of the Great Lakes under Lake-Breeze Conditions

Lake breeze pollution comes inland, then rises in the return flow.

Cole and Lyons have suggested that transportation planners should carefully consider these lake effects. Concentrations of freeways and highways near shore areas may create serious pollution problems resulting from automobile emissions which are not effectively dispersed. The occurrence of a photochemical smog problem similar to that of Los Angeles, California, is not impossible. Given the high values of sunshine along the lakeshores in summer (due partly to the suppression of convective cloudiness by the lake), poor dispersal of NO_x and hydrocarbons from automobiles combined with increased automobile traffic could cause serious pollution problems of a photochemical nature.

Problems could also arise following the construction of airports along the immediate lakeshore or on polders in the lake, as proposed in the Chicago area. Pollutants might be carried into the city by the stable lake breeze or onshore flow of air during the summer, adding to pollution levels which are already high.

In the future industrial and urban development of the Great Lakes, the special problems posed by the unique meteorological features of the Lakes cannot be ignored. Remedies are difficult and expensive. Preventive measures are much more effective and, in the long run, less costly.

16: Deliberate Modification of Weather

SINCE THE BEGINNING OF HISTORY, MAN HAS CONtemplated various means of changing the weather. Incantations to the deities, ceremonial dances and rituals, and primitive rockets launched into the air were some of the earlier weather modification attempts. Some efforts apparently met with success, or at least so it was thought. For example, the ancient Greeks felt they had been able to improve the climate of the Aegean. They had importuned Aristaeus, the son of Apollo and Cyrene, to ease the summer heat which burnt the Aegean. Aristaeus, it was believed, built an altar in honor of Zeus, and Zeus answered by sending the etesians, winds which blow during the summer over the Aegean with great constancy, to cool and refresh the islands.

While the rainmaker and the rain dance remain as reminders of man's earlier efforts to alter the weather, the era of scientific weather modification awaited the acquisition of more precise knowledge of the atmosphere and did not really begin until nearly the middle of the twentieth century. In 1891, a scientist had an idea that dry ice could be used to modify clouds, but it was a fortuitous event in 1946 which proved to the world that this was actually possible. During that summer, Vincent

Schaefer, a scientist for General Electric, was experimenting in Schenectady, New York, with a small food freezer in an attempt to study supercooled clouds. (Supercooled clouds are those which are composed of small liquid water droplets, even though the temperatures may be considerably below freezing.) The chance event which occurred during one of these experimentations is best described in Schaefer's own words:

> During the late spring of 1946, I tested a variety of chemicals selected from our well-stocked shelves, dusting them into the chamber with and without supercooled cloud. I didn't have any appreciable success. Then early in July we had a period of hot weather with our laboratory becoming so warm as to affect the cooling capacity of my small chamber. I decided to use some dry ice to assist in cooling the chamber. The instant I put the dry ice into the supercooled cloud, a spectacular change occurred in the supercooled cloud and I knew that I had achieved the transformation I had been seeking. (Schaefer 1968)

The change which had occurred was the almost immediate production of a myriad of tiny ice crystals. Subsequent experimentation that year indicated that significant alterations of supercooled clouds followed the introduction of dry ice or liquid carbon dioxide. It was also found that cloud alterations could be produced by introducing silver iodide into the cloud.

The Theory of Cloud Seeding

These events in 1946 heralded the beginning of a long series of cloud-seeding experiments and ushered in the modern era of weather modification. Deliberate attempts by man to alter the weather include rain making, rain suppression, fog dispersal, cloud dispersal, hurricane suppression, hail suppression, and snow redistribution. All of these rely upon cloud seeding to produce the desired effect; consequently, a wide spectrum of possible modifications stem from seeding processes. Although some forms of weather modification may rely on techniques other than cloud seeding, seeding is basic to so many modification possibilities that it is important to understand its implications and limitations.

Clouds, as mentioned, consist of tiny water droplets or ice crystals. These form when moisture condenses (or crystallizes) on tiny nuclei. The nuclei for water droplets are called *cloud condensation nuclei* (*CCN*) and are abundantly available in the atmosphere. Supplied largely from natural sources, they also inadvertently result from human activities. The nuclei for crystal formation (called *ice-forming nuclei* or *IFN*) are not quite so abundant and, in fact, may sometimes be rather scarce. A rather limited variety of substances may activate the formation and subsequent growth of ice crystals. Like CCN, though, some substances which are good ice-forming nuclei may be given off inadvertently into the atmosphere. Near steel-producing areas, the supply of IFN seems unusually large.

It is important to recognize that a cloud may consist of both liquid water droplets and ice crystals. Water droplets can retain their liquid state at subfreezing temperatures if the droplets are very small. The average radius of liquid cloud droplets is only about 10 microns (remember, a micron is 1/10,000 of a centimeter), so the water droplets which make up clouds are indeed tiny—so small that they are literally suspended within the cloud. Also, the volume of an average-sized cloud droplet is so small that the cloud droplet need move only a very short distance before being consumed by evaporation.

Precipitation from clouds

With a cloud consisting of average-size cloud droplets, it can easily be seen that some major impediments must be overcome before these droplets may reach the surface as precipitation. Size and weight of the droplets must be increased greatly (to raindrop size) so they will no longer be suspended (terminal velocities must be increased) and may thus fall to the surface without being evaporated. Unless this growth occurs, clouds will not give precipitation.

The diameter of an average raindrop is several hundred times that of the average cloud droplet, and its volume is about one million times that of the cloud droplet. Essentially, two ways exist by which cloud droplets may be induced to grow to the much larger raindrop size. One way results when the cloud droplets bump together or collide. This growth process is known as *coalescence* and may result in warm rain.* For coalescence to occur, it is necessary for some differences to exist in the sizes of the original cloud droplets so that differences in fall velocities may occur. In that way, larger cloud droplets may overtake smaller cloud droplets, and their water volumes may join when a collision takes place. Fairly wide differences in cloud droplet sizes have been noted over oceans, particularly tropical oceans. However, over continents the size differences of the cloud droplets are smaller. Thus the coalescence process is likely to go on with comparative ease over the tropical oceans but with difficulty over the continents.

Crystal growth may occur when crystals and supercooled water coexist within a cloud. When this situation occurs, the supercooled water droplets evaporate (because saturation vapor pressures over water at subfreezing temperatures are larger than those over ice) and the moisture diffuses to the surface of the crystal, causing the crystal to grow. As the crystal grows, it becomes larger and heavier and begins to fall through the cloud. It may collide with other crystals and grow further (this process is called *clumping*), or it may collide with supercooled water droplets which immediately freeze on the crystal (*riming*). Both clumping and riming cause the crystal to increase in weight and size. In addition, the crystal may also grow through the

*Warm rain is so called because scientists noted that rain often fell in tropical areas from clouds with above-freezing temperatures throughout.

continued diffusion of moisture as supercooled water droplets evaporate and the water vapor moves to the surface of the crystal.

Over the Great Lakes region, much of the precipitation is probably a result of the mixed presence of supercooled water droplets and ice crystals in clouds at subfreezing temperatures. Even during the heat of summer, the freezing level may be reached at heights of 12,000 or 13,000 feet (3,650–3,950 meters), and many clouds extend much higher than that. Once the ice crystals enlarge through the processes described, they may fall to lower levels where they may melt. The clumped and rimed crystals (snowflakes) then become large raindrops which can grow further by the coalescence process. This means of precipitation is called the *cold rain* process. The only catch to the whole sequence of steps is that, as we have said, ice crystals lack abundant nuclei around which to form. Thus their abundant formation within the cloud is rather difficult. But suppose suitable nuclei could be deliberately injected into the cloud or the crystals themselves could be introduced into the cloud. Wouldn't this affect the capacity of the cloud to give precipitation? It would certainly seem so. This is where cloud seeding enters the picture.

By introduction of silver iodide into the cloud, either from ground-based or airborne generators, large numbers of crystals can be formed. These crystals attract moisture to their surfaces, thus causing them to grow and become heavier. Eventually rain or snow is received. Simple? Seems so. Why, then, have 30 years elapsed with only slow progress made toward regular operational employment of cloud-seeding methods?

Some problems of cloud seeding

Let's consider some hypothetical situations and attempt to determine theoretically what results might follow seeding. In the first situation (figure 81A) a thick cloud which contains a large

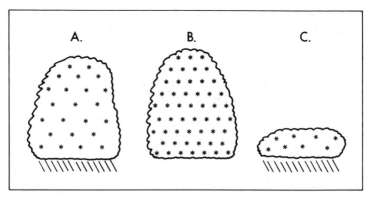

A. Seeding to enhance precipitation. Crystals become larger, heavier, and fall from the cloud. Precipitation is augmented.
B. Overseeding. Many small crystals are competing for available moisture; so none grows large enough to fall from cloud. Precipitation inhibited.
C. Seeding of a thin, supercooled cloud. Crystals grow and fall from the cloud as a brief flurry of precipitation; then the cloud dissipates.

Figure 81   Some Possible Results of Cloud Seeding

quantity of moisture is selected. We will seed the cloud with a measured amount of silver iodide. Quickly a large number of crystals will form within the cloud. The generous moisture supply available will cause each crystal to grow and fall out of the cloud as precipitation. We have successfully increased the precipitation which might have normally occurred by seeding in just the right amount.

However, suppose the cloud is overseeded (figure 81B). Too much silver iodide is supplied, and so many crystals are produced that competition for the available moisture is intense. Each crystal grows a little, but none becomes large and heavy enough to fall from the cloud. In this case, precipitation may have actually been suppressed by overseeding. The question is, what is the right amount of seeding? And since clouds are extremely variable in vertical thickness, moisture content, and temperatures, how will we obtain the information we need about the cloud in order to decide how much to seed?

One more hypothetical situation may show some of the additional problems encountered when one attempts to seed clouds. Suppose we choose to seed a cloud which is fairly thin and contains a limited amount of moisture (figure 81C). With seeding, many crystals are formed which immediately attract moisture and grow, falling out of the cloud as rain or snow. But in a cloud with limited moisture supply, a brief rain or snow shower may be all that can be produced; then, because the moisture is used up, the cloud actually dissipates. So by seeding different types of clouds with differing amounts of seeding agent, we may be able to increase precipitation, decrease precipitation, or actually dissipate the cloud.

Much of the early optimism over the possibilities of cloud seeding reflected the fact that scientists actually did not know much about the way in which precipitation naturally occurs within a cloud. For cloud seeding simply tries to hasten the processes which would naturally occur to cause rain or snow. It was back to the drawing board for the scientists, and a long period of basic study of cloud physics ensued.

Meteorologists today know much more about what goes on inside clouds but are certainly far from having the ultimate answers. Cloud-seeding operations are fraught with difficulties. In addition to varying moisture contents, the temperature structure of the cloud plays a role in determining the mode of crystal formation and the ease with which crystals will form. A major problem also arises when the crystals grow. The heat released by the freezing of water (*heat of fusion*) is released into the air in the cloud. This, in turn, alters the temperature and affects the buoyancy of the cloud. In fact, seeding of clouds with strong updrafts and stimulating them to grow (*dynamic seeding*) may be the primary basis for increasing rainfall from some types of clouds. The lifetime of the cloud and the sizes and concentrations of cloud droplets and ice particles are other important variables.

Problems exist concerning where the plume of silver iodide released by the generator actually goes. Does it really get into the cloud? And at what level? Major problems also exist in attempting to evaluate cloud-seeding experiments and operations. How is it determined whether the precipitation following a cloud-seeding venture would or would not have occurred without the cloud seeding? And does rain making involve "robbing Peter to pay Paul"? Is precipitation increased in one area at the expense of decrease in another?

Weather Modification Today

Currently, the role of weather modification in offsetting unfavorable natural weather features is being looked at very closely. The capacity to make rain, to break up clouds, to dissipate fogs, and to prevent lightning occurrences may well allow us to forestall weather crises. The American Meteorological Society, in an attempt to appraise the present status of weather modification, has recognized that under certain conditions proven results may be realized following cloud seeding. With cold orographic clouds such as those which form along the western slopes of the Rockies in wintertime, increases of precipitation may occur as a result of seeding. Additional snowpack may be very desirable, as the snowpack in the higher mountains provides much-needed irrigation water during the growing season. Also, cold fogs (those composed of supercooled water droplets) may be dissipated by seeding. This may be very desirable when airports are closed due to low visibility. (Unfortunately, many if not most fogs are warm fogs which cannot be seeded. Researchers are currently attempting to learn more about the possibilities of dissipating warm fog.) Progress has also been made in suppressing hail and lightning and in seeding of hurricanes to lessen their intensity. In spite of these recognized achievements, there is still a very long way to go before cloud seeding and weather modification will be commonly employed methods of avoiding unfavorable weather events. For small but intense storms, such as severe thunderstorms and tornadoes, the future does not appear very hopeful. Meteorologists must learn a great deal more about how these storms form and how they may be forecasted before successful modification procedures can be employed.

Weather modifications in the Great Lakes area

What are the possibilities of modifying the weather in the Great Lakes area? Cloud seeding is certainly no stranger to the region. Private cloud-seeding concerns have been operating within the Great Lakes basin for a number of years. Their goal is chiefly to augment the rainfall during the agricultural season. Attempts such as these occur in many parts of the country. But possibilities of modifying weather and climate in the Great Lakes area also relate to the unique lake effects. It is in this realm that we may see a concentration of weather-changing attempts during the next several decades.

Let's take the example of lake-effect snowfall. It has been estimated that the moisture from a typical lake-effect snowfall may provide a 16-day water supply for the entire twenty-eight million persons who derive their fresh water from the Great Lakes hydrologic system (Weickmann 1972). This water can then become a very valuable commodity. Suppose, however, that the lake-effect storm extends far enough inland so that the snow is received *outside* the basin. This is possible, particularly in New York State, where the basin and watershed extend only a small distance inland. In this case, the water is lost from the Lakes system. This may be desirable during periods of high lake levels, but when the lake levels are low it represents the loss of a valuable asset.

Mathematical models have been developed which indicate that, by seeding, the snowfall could be induced earlier, over the lake instead of inland. (Overseeding, however, might cause the snow to fall farther inland [Lavoie, Cotton and Hovermale 1970].) By keeping the water within the Great Lakes basin, an addition to the supply of fresh water is made. Seeding over the Lakes may increase as well as redistribute the precipitation. Feasibility studies (Stout and Ackerman 1974) have concluded that increases of fall precipitation over Lake Michigan on the order of 10% to 20% could result from seeding of unstable season clouds. This increase would be felt in the downstream lakes in terms of an increase in lake levels. Obviously, this increase of lake-water volume and level might not be viewed with favor by all. It might become a benefit during low-water periods to power generation and navigation but would also be some problem because of the added shoreline erosion resulting from the higher lake levels. With a goal of keeping the Lakes near their long-term average levels, cloud seeding could be used during low-water periods.

There are also other ways by which the weather and climate of the Great Lakes might be changed. The low clouds which linger around the Lakes in winter make that season very gloomy; with more sunshine, the long winters might be a bit more tolerable. Very often these clouds are rather shallow and are supercooled. The addition of ice-forming nuclei may result in brief snow showers followed by dissipation of the cloud. Trial runs near Alpena, Michigan, dissipated about 175 square kilometers of a supercooled cloud deck by cloud seeding and a field program in February 1977 further investigated the feasibility of dissipating supercooled clouds. Could this be done on a regular operational basis? Possibly. It would undoubtedly be costly; but the major problem is this: Just who would decide to do it? And to whom would it be beneficial?

Weather modifications and the inhabitants of the Great Lakes basin

The weather modification possibilities in the Great Lakes basin present ample opportunities for social and political conflict. If lake-effect snows were partially redistributed back over the lake, how would ski resort operators, whose living depends on frequent lake-effect snow, respond? Certainly, not with smiles if they thought they hadn't received enough snow. If cloud seeding is used to raise the level of the Lakes during low-water periods, what redress would be available for the lakeshore property owner whose beach has eroded? Suppose a lake-effect snowstorm, which normally would deposit most of its moisture over the hills and countryside of southwestern New York State, is seeded to redistribute the snow closer to the lake but a small error of estimation results in the city of Buffalo being buried! Transportation, businesses, schools, homes, people: All are deluged by a snowfall which they would not have received except for man's deliberate interference with nature. What redress is available? Or do the citizens of Buffalo have to grin and bear it?

The possible conflicts of interest and the difficulty of determining just what is beneficial to whom point out the fact that our social and political institutions are not yet prepared to deal with the intricacies of weather modification. Weather modifiers have for a long time been viewed with suspicion by the public, partly because they have operated within a veil of semisecrecy. Within the past several years, the public has become more actively involved in deciding whether modification activities will or will not occur. However, the record is not free of skirmishes between the weather modifers and those who have suffered damage, either real or imagined.

A number of controversies have been generated as a result of damage claims brought by groups or individuals. For example, cloud-seeding operations were proceeding in the Black Hills area just prior to the infamous flood of June 1972. Some people as well as the news media blamed the cloud seeding for the flood, although this was denied by scientific investigation teams. Other public controversies have developed in Colorado and Florida as a result of weather modification. A lawsuit in Michigan and an intense controversy in Wisconsin during the summer of 1977 over whether cloud seeding should occur are recent examples of social and legal questions raised in the Great Lakes area. These controversies have pointed to the need for greater public participation in weather modification decision making and for a legal framework within which modifiers must operate. The drafting of legislation dealing with weather modification has been primarily the job of the states. The federal government has required only that the activity be reported. This is chiefly to preclude the possibility of contamination from a nearby seeding operation of which the modifier was unaware.

As of 1977, 31 states had weather modification laws, but only one of these states was located in the Great Lakes area. Some of the laws were favorable, some unfavorable, to the technology of weather modification. Many other states, however, had recognized the desirability of weather modification legislation; and some, including Michigan, had laws under consideration by their legislatures. In the Great Lakes region, the state of Illinois had developed what might be considered a model piece of legislation.

What factors should a good weather modification law consider? A declaration of purpose should introduce the document. The purpose of the Illinois law, and the accompanying statement, read as follows:

> Weather modification affects the public health, safety and welfare and the environment, and is subject to regulation and control in the public interest. Properly conducted weather modification operations can improve water quality and quantity, reduce losses from weather hazards, and provide economic benefits for the people of the State. Therefore, weather modification operations and research and developments shall be encouraged. In order to minimize possible adverse effects, weather modification activities shall be carried on with proper safeguards, and accurate information concerning such activities shall be recorded and reported to the department. (Nurnberger 1974)

A responsible authority or agency is then normally designated as the administrative agency. In Illinois, a five-member Weather Modification Board was established within the Department of Registration and Education. Members of the board are appointed and include individuals with qualifications and practical experience in agriculture, law, meteorology, and water resources.

The various duties of the designated agency are then outlined. These include the licensing and permit-granting procedure, the acquisition of materials, equipment, and facilities, the receiving of funds, the making of studies, the conduct of research, the representation of the state in interstate compacts, and so on. Procedure for hearings and exemptions from permit requirements may also be considered. Provisions regarding the suspension or revocation of a license are also established, as well as liability considerations, with procedures for redress.

Whatever the future of weather modification in the Great Lakes area (and the apparent, technologically feasible opportunities for changing the weather are numerous), greater public participation, both in decision making and in financing, is likely. Such participation will also require a restructuring of some social, political, and legal systems. More and more states and provinces are recognizing the desirability of laws to regulate weather modification activities and to assist in modification ventures. This means that citizens in the Great Lakes area must become aware of the capabilities and limitations of current weather modification technology.

PART FIVE

Great Storms of the Great Lakes

17: The Big Storms

> . . . For in their interflowing aggregate, those grand
> fresh-water seas of ours—Erie, and Ontario, and Huron,
> and Superior, and Michigan—possess an ocean-like ex-
> pansiveness, with many of the ocean's noblest traits; with
> many of its rimmed varieties of races and of climes. . . .
> They are swept by Borean and dismasting blasts as direful
> as any that lash the salted wave; they know what ship-
> wrecks are, for out of sight of land, however inland, they
> have drowned full many a midnight ship with all its
> shrieking crew.
>
> Herman Melville, *Moby Dick*

ON 25 MAY 1976, UNDERWATER CAMERAS LOWERED
in 530 feet of water to the bottom of Lake Superior's Whitefish
Bay returned photographs of the hull of the ore carrier *Edmund
Fitzgerald,* broken into three pieces. She was the latest of the
commercial vessels plying the Lakes to become a weather-
related casualty.

Built in 1958, she was at that time the largest on the Great
Lakes. The gargantuan vessel was designed to haul hematite,
and later taconite, from the Mesabi range in northeastern
Minnesota to the blast furnaces which line the shores of the
lower Great Lakes.

When she left the Burlington Northern docks at Superior, Wisconsin, during the early afternoon of Sunday, 9 November 1975, winds were light and weather was warm by Lake Superior standards. After a chilly September, a mild fall had set in, and the usual succession of autumn gales on the Lakes had been absent. It was an autumn which would have been remembered chiefly for its abundance of balmy, sunny days if not for a powerful November storm which was to send the *Fitzgerald* to her demise in Whitefish Bay.

On Saturday, 8 November, forecasters noted that the barometric pressure was beginning to fall over western Texas. That evening a weak low-pressure area was evident on the surface weather map, and the beginnings of a storm circulation could be detected. At the same time, cold air began pushing southward over the mountain states and warm air began to advance northward from the Gulf of Mexico. A telltale counterclockwise wind pattern could be discerned around the center of the newly born cyclone. This weather system would bear watching, as forecasters have long been aware that the high plains area is a favorite region for cyclone development—or, in the technical terminology of the meteorologist, *cyclogenesis*. They knew that storms forming in the high plains of Texas and Oklahoma tend to intensify quickly, to become severe, and to track northeastward to the Great Lakes or New England.

By Sunday morning, the cyclone had moved to southern Kansas with continued falling pressure. Warm air from the south and cold air from the north now began to collide vigorously near the center of the system. Potential energy, made available when portions of the atmosphere with unlike temperatures collide, became converted to the kinetic energy of motion—the winds. Energy far greater than that released by the most powerful of nuclear explosives was poised to fuel the developing weather system.

As yet, on Lake Superior, there were few foreboding weather signs. A routine downlake trip seemed to be in store for the *Fitzgerald,* her crew of twenty-eight, one cadet seaman, and a cargo of 26,000 tons of taconite pellets. On Sunday evening, however, an almost imperceptible shift of wind to the northeast occurred. The center of the developing storm was now over eastern Iowa, intensifying rapidly. At the higher levels, a favorably aligned jet stream was assisting in the rapid conversion of energy needed to sustain the storm and was also providing a steering mechanism which would guide the storm directly over Lake Superior toward a head-on encounter with the *Fitzgerald.*

By Monday morning, 10 November, the central pressure of the storm had fallen to 982 millibars, extraordinarily low for a storm of other than tropical origin. It was centered near Marquette, Michigan, with gale-force winds over the eastern end of Lake Superior which slowly veered to the east, then to the southeast as the storm continued to track northeastward toward the mountainous northern shore of Lake Superior. The *Fitzgerald* had taken the northern route, hugging the Canadian shore, and was headed for Whitefish Bay and the locks of Sault Ste. Marie. The storm center, a vortex of a vast whirlwind, passed almost directly over the *Fitzgerald* during the day, with central pressures continuing to fall as the storm reached Canadian waters and headed toward the Ontario shoreline. Ironically, the *Fitzgerald* weathered the storm's frontal assault. She plodded steadily eastward, taking water under heavy seas stirred by winds of near hurricane force, but apparently was in no immediate danger.

It was the tail end of the storm which struck the death blow for the *Fitzgerald.* By Monday evening, the cyclone was centered near Moosonee, Ontario, moving toward northern Labrador where it would eventually decay and dissipate. On its southern flank, it readied yet another onslaught for the *Fitzgerald.* Winds on eastern Lake Superior veered from southeast to south, then into the southwest and west, where they attained peak velocities. It was a final gesture of defiance as the storm expended itself over the barren tundra of northern Labrador; and by morning, the gales which had swept Lake Superior for 24 hours were gone. So also was the *Fitzgerald*.

The steamer *Arthur M. Andersen,* owned by the U. S. Steel Company, was following the *Fitzgerald* as she turned southeastward off Whitefish Point toward Sault Ste. Marie. Radio contact moments before the *Fitzgerald* vanished indicated that although she was listing slightly and taking some water, no serious problems were being encountered. The master of the *Andersen* stated that his last contact with the *Fitzgerald* was by radar. Shortly after 7:00 P.M., she vanished from the radar screen. By Tuesday morning, news of the *Fitzgerald's* disappearance had been reported nationwide.

A report issued nearly 2 years later by a Coast Guard board of inquiry indicated that progressive flooding of the *Fitzgerald's* hold had caused the ship to ride lower in the water. With continued leaking into her hold, she suddenly began to sink, reaching the bottom with the screw still turning. The sinking was so sudden that no attempt was made to abandon ship. The Lake Carriers Association rejected this theory, insisting that the *Fitzgerald* sank after hitting a shoal, or underwater reef.

Over 300 years before the foundering of the *Fitzgerald*, the first commercial vessel on the Great Lakes, the *Griffin,* suffered a similar fate somewhere in the storm-swept waters on northern Lake Huron, then called Lake Dauphin. There was little publicity and investigative inquiry of the sort which followed the *Fitzgerald* disaster. Father Hennepin wrote some years later that it could never be learned what course the *Griffin* took nor the circumstances under which the vessel had perished.

In the years between the disappearance of the *Griffin* and the loss of the *Fitzgerald*, the weather of the Great Lakes has claimed a steady toll of vessels and humans. Mystery shrouds the circumstances surrounding the loss of many of the ships; and names like the *Rouse Simmons,* which disappeared in the autumn of 1912 enroute from Manistique to Chicago loaded with Christmas trees, the *Regina,* and the *Price,* both lost during the terrible 9 November 1913 storm on Lake Huron— and, more recently, the *Fitzgerald*—have become entwined with supposition and legend.

Each year has its quota of storms, but a few stand apart and join the lore of the Lakes. These great storms are imprinted indelibly in the minds of lakefaring men. They take their toll of shipwrecks, spawn tales of heroism, and inspire books.* The agony of the freighter *Mataafa,* witnessed by thousands as she lay writhing in Duluth harbor under the thrashings of one of the great storms of the century is relived in numerous writings. The *Charles S. Price,* ambushed by one of the most unusual storms ever to hit Lake Huron; the Canadian pulpwood carrier *Novadoc,* along with the *William B. Davock* and the *Anna C. Minch,* all victims of the havoc on northern Lake Michigan on Armistice Day 1940; the *Carl Bradley,* a more recent victim of a furious storm—these are names of ships immortalized by the big blows on the Lakes. Common to many of the shipping disasters was a reaction of awe and disbelief that such miracles of modern marine engineering could fall victim to the winds and waves of inland seas. But as Melville had warned, they were capable of unleashing all the untempered fury of oceans.

In fact, they may occasionally become even more dangerous. On the high seas, captains have more room to navigate their ships. Vessels can be tossed to and fro and driven far off their course without the ever-present danger of being driven ashore and having a hold ripped asunder by the jagged rocks of rugged coastlines. Vessel masters prefer Lake Superior if they are forced to ride out a storm because there is more room to navigate. Lake Michigan is feared. With its lack of natural harbors and protective shelter, along with its tricky currents, it can be a devil of a place to sail in a storm. More ships have been

---

*The great storm of 1913, for example, is the subject of the book *Freshwater Fury* by Frank Barcus (Wayne State University Press, 1960).

lost on Lake Michigan than on all the other Lakes combined. Its elongated dimensions make it a perilous body of water when the wind blows strongly from the north or south. Lake Erie is also dangerous; being shallow, waves can be easily generated. Sudden storms and squalls, which may not persist long enough to form large waves on the other Lakes, can do so on Lake Erie.

Even the waves on the Great Lakes seem different from those of the open seas. Ship captains have observed that on the Lakes, with their restricted fetches, waves seem to have a different type of motion. They appear to come at more rapid intervals than ocean waves, to "jump and tumble rather than roll and swell" (Ratigan 1960, p. 15).

Great Lakes shipping in the early days of weather forecasting

Consider the plight of the ship captains of the 1800s and early 1900s. The U.S. Weather Bureau at that time was a neophyte agency and forecasting had not yet breached the threshold between art and science. Little atmospheric data became available other than from a thin slice of air next to the surface. Jet streams were unheard of, and their influence on the forming and steering of powerful storms was not even dreamed of. The Norwegians had not yet given the world the concepts of fronts and air masses.

The Weather Bureau forecast was regarded with suspicion by the lake men. A preexisting storm which appeared to be headed toward the Lakes could be forecast with some degree of accuracy, but the ability of the Weather Bureau to predict the development of new storms or to cope with the occasional unusual weather situation was severely limited. There were no computers and sophisticated mathematical models, and weather forecasts missed a lot more than they do now.

Given this state of the art of forecasting, the ship captains preferred to rely on their own experience, judgment, and intuition. Their years of experience on the Lakes had imbued in them a certain sense of anticipation of future weather which they relied on but could not explain scientifically. In his book *Great Lakes Shipwrecks and Survivals,* William Ratigan succinctly describes the approach of the average ship captain to weather forecasting in the early days:

> There were captains on the Lakes who had no more thought of carrying a barometer than today's car driver thinks of an extra accessory such as radar. A typical skipper wet his thumb to tell which way the wind was blowing and he relied on such optimistic weather forecasts as "evening red and morning gray will set the sailor on his way" or "rainbow at night, sailors delight," but he ignored all the pessimistic jingles. (Ratigan 1960, p. 95)

Often the skipper's approach to weather prediction worked, but occasionally it did not and the results were disastrous.

Today, giant computers feeding on data from an intricate grid system of surface and upper air stations grind out daily prognosis charts indicating conditions from 12 to 72 hours in the future. While the interpretation of these machine-made predictions for the exact weather conditions of specific areas still requires all the skill, experience, and knowledge that the forecaster in the Weather Service Forecast Office can muster, there seems little likelihood that a major storm could sneak into the area undetected. Radio and ship-to-shore telephone keep the ship captains in constant touch with weather information, and their radars scan the horizons endlessly for the presence of obstacles and forbidding shorelines. Ships of modern design with sophisticated guidance systems ride through tempests with an ease which masters of their ancestral vessels would envy. Yet the most amazing thing of all is that the storms still take their toll.

Features common to the big storms

Invariably the big storms occur in November, which has

always been a stormy month on the Great Lakes. At this time of year the atmospheric circulation quickens and large temperature contrasts exist between the vast northern expanses of Canada, already snowcovered, and the lingering warmth of the Gulf states. The ships are still on the Lakes and, with the shipping season rapidly drawing to a close, the captains are eager to deliver that last grain shipment or taconite cargo. Sometimes in the past, chances were taken when ordinarily better judgment would prevail.

With many of the great storms, a misinterpretation of weather signs leading to fateful decision making was a big factor in the toll of vessels and men claimed by the Lakes. Sudden and unprecedented wind shifts exposed the vessels to the full fury of winds and drove them close to menacing coast lines.

Close cousins to the fierce storms of the early days can be found during more recent decades. For example, the big storm of 1950 behaved in a very similar fashion to the great storm of 1913. And some striking similarities existed between the storm which caused the breakup of the *Fitzgerald* and the Armistice Day storm of 1940. With the more sophisticated data gathering facilities and forecast methods of the later eras, analytical hindsight could be used to study the behavior of the earlier storms.

Wherever and whenever men of the Lakes gather to reminisce over their sailing days, two storms stand out as the most savage, destructive, and memorable. They are the 9–10 November 1913 storm on Lake Huron and the Armistice Day storm of 1940 on Lake Michigan. The storm of 1913 was by far the most destructive, but many lake men will argue that the 1940 storm was more powerful. Each of the storms was a very intense low-pressure system which came out of the south, and each was accompanied by strong winds and heavy snows. In spite of these likenesses, each was unique, a once-in-a-generation happening.

The Great Storm of 1913

It might have been a premonition, or perhaps fatigue after the long sailing season and extended absence from home and family. But whatever his reason, First Engineer Milton Smith of the *Charles S. Price* did not reboard his ship on Saturday, 8 November 1913. The *Price* had left Ashtabula, Ohio, with a full load of coal and was heading up the Lakes on one of her last voyages of the year. Smith got off at Cleveland and could not be persuaded to return. A few days later, he was identifying the bodies of his shipmates, victims of one of the most savage and undoubtedly the most destructive of all the Great Lakes storms.

Lake Huron was hardest hit. Lake Michigan escaped unscathed, with not a vessel lost. The western end of Lake Superior was also spared most of the storm's impact, but several vessels were lost in eastern Lake Superior. On Lake Huron, eight vessels foundered with their entire crews. The death toll rose to over 250. Over fifty vessels were severely damaged, and losses ran into the millions of dollars.

Some of the ships which succumbed to the storm were new and modern steel bulk carriers. The *James Carruthers* exemplified the extra safety measures which had been poured into vessel construction during the early part of the century. She was reinforced by extra steel in her hull and was nearly twice the length of a football field. Like the reaction after the breakup of the *Edmund Fitzgerald,* it seemed impossible that she could be among the missing.

The Weather Bureau drew criticism from some quarters, chiefly for its failure to predict the quick wind shift which had caught the vessels by surprise. But the Bureau could hardly be blamed. The storm was a rarity, a once-in-fifty-year occurrence which surprised even the wisest, most experienced forecasters. At that time it was never fully understood; and it was not until 35 years later that a similar type of storm, alike in birthplace and life history but fortunately lacking the powerful winds on the

Lakes which accompanied the 1913 debacle, revealed its true character to the weathermen.

Early warnings

Strangely enough, the Weather Bureau had correctly issued storm warnings throughout the Great Lakes but for the wrong reason. On Friday, 7 November, a low-pressure area began to drift across the upper Lakes, reaching northern Lake Huron on the morning of 8 November. There was nothing particularly unusual about this, as low-pressure areas are common in November when the Alberta cyclone track becomes active.

The Weather Bureau dutifully tracked the storm eastward from northern Minnesota and issued the following warning: "Hoist southwest storm warning 10:00 A.M. Storm over upper Mississippi valley moving northeast. Brisk to high southwest winds this afternoon and tonight, shifting to northwest Saturday on the upper lakes" (Barcus 1960). Some rough weather loomed ahead, but nothing unprecedented. Skippers knew that winds veering from southwest to northwest normally signaled the continued movement of a cyclone eastward and its progression through Canada. Its tail end would lash the Lakes for some hours, but with the shift of the winds to the northwest, the gales could be expected to blow themselves out—no great concern, but best to stay close to the protective shores. All should be well by Sunday.

On Saturday 8 November, the Weather Bureau bulletin read: "Change to northwest storm warning 10:00 A.M. Storm over eastern Lake Superior moving east-northeast" (Barcus 1960). The pressure near the center of the storm had dropped to 1,000 millibars (29.53 inches of mercury)—rather a low barometer for this time of year, but still nothing to get excited about. Up to this time, the Weather Bureau had issued entirely appropriate and adequate warnings.

But now the unexpected occurred. A quick switch of the

wind to the northeast on Lake Huron was accompanied by an increase to hurricane strength. Any mariner on the Lakes worth his license knew that "noreasters" are unlikely to occur at the tail end of a "norwester." The steamer *James S. Dunham,* heading out of Duluth, Minnesota, late Friday, had been prepared for the high northwest winds by the Duluth weather station, but not for a northeast storm. Captain Frank Pratt blamed the Weather Bureau for supplying inadequate warnings of the wind shift.

Starting from the other end of the Lakes at Buffalo, New York, Captain Watson of the *George F. Brownell* described the weather conditions along his route. The winds on Lake Erie, just as the Weather Bureau had predicted, held southwest and west, with a low barometer. At 8:10 P.M. Saturday, 8 November, the *Brownell* was at the Lake Huron lightship at the extreme southern end of Lake Huron. Watson described the wind as "west and fresh, the clouds very low, of whitish color and hard looking" (Barcus 1960, p. 118). By 12:40 A.M. Sunday, 9 November, the *Brownell* was off Harbor Beach, Michigan, with a fresh west-northwest wind. At 2:30 A.M. came the unprecedented wind shift into the north and northeast. A quick increase in velocity was followed several hours later by snow and, throughout the day, increasing seas. The stage was set for one of the worst storms ever on the Lakes.

The *Brownell,* more fortunate than others, made it through the Straits of Mackinac and worked her way down the protective west shore of Lake Michigan. In the meantime, the *Charles S. Price* had "turned turtle" in southern Lake Huron, with all hands lost. She was floating upside down just a few miles off shore.

The origins of the great storm of 1913— a visitor from Virginia

The confusion over the storm's nature and origin extended

to both the Weather Bureau personnel and the vessel skippers. The Weather Bureau defended itself by noting that storm warnings had been flying all along the Lakes almost 48 hours before the shift of winds which signaled the beginning of that fateful Sunday. To be sure, the Bureau had predicted a storm but had no inkling of the veering of the wind to the northeast and the ferocity of the winds which would follow. The Bureau had been watching the percursor storm over the upper Lakes, which in reality was quite blameless for the havoc of that November weekend. A new and unforeseen visitor from the south, tracking an amazing route, had defied all the rules of meteorology and approached the Great Lakes from the *southeast*!

At the time, the origin and nature of the new storm went almost completely unrecognized, and most lake men felt that the precursor storm over the northern Lakes had somehow intensified or retracked to strike the Lakes another blow. Struggling to explain a weather situation which had never before confronted him, Captain Watson of the *Brownell* stated: "I think the Storm had been traveling along the regular storm track and after reaching the Great Lakes region was diverted from that track so suddenly that its deflection couldn't be noted until after the storm was upon us. The high north-northeast winds were perhaps a sort of flare back of the high southwest winds of the day previous" (Barcus 1960, p 117).

What had happened? Why the sudden shift of winds? The weather maps held the answer. The events of 1913 have been reconstructed via surface weather charts to show the unparalleled formation of a secondary storm over the Carolinas and Virginia and its subsequent movement *northwestward* into the Great Lakes (figure 82).

On the morning of Saturday, 8 November, the low-pressure area over the upper Lakes, for which the Bureau had issued storm warnings, was accompanied by a front which extended southeastward to western Pennsylvania, thence southward to northern Georgia. With the normal progression of weather

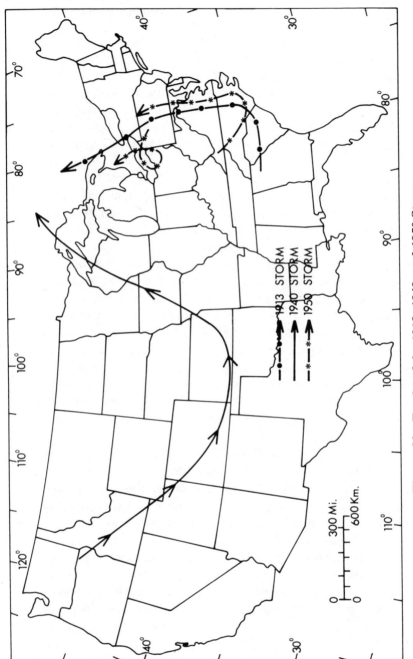

Figure 82 Tracks of the 1913, 1940, and 1950 Storms

1913 STORM
1940 STORM
1950 STORM

events which would be expected to accompany such a weather situation, the storm and its attendant front would simply move on east, out of the area.

The key to the birth of the anomalous storm was the development of a small wave in northern Georgia along the attendant front and, subsequently, a secondary cyclone (figure 83). By Saturday evening, the newly developing low was over South Carolina and intensifying rapidly. The old low-pressure area over the upper Lakes was weakening now and was no longer a real factor. Captain Watson on the *Brownell* was to note in his log that as he entered the St. Clair River about 4:00 P.M. on Saturday, the wind seemed to decrease in velocity although it was still fresh from the west and southwest.

As the upper Lakes cyclone slowly weakened, the nascent storm moved north-northwestward to Virginia, where it was located Sunday morning, and continued to intensify, with central pressure at 990 millibars (29.24 inches of mercury). The Weather Bureau noted this in its morning bulletin for 9 November: "Continue northwest storm warning. Storm over Virginia moving northeast" (Barcus 1960). But there was little to suggest that the Virginia development would create problems for any areas other than the eastern seaboard and New England.

Although the forecasters had no way of knowing, a powerful flow of air aloft had developed into a most unusual configuration. The jet stream swept southward over the central Midwest to northern Georgia. From there it made almost a 180-degree turn and careened northward to the eastern Lakes. It not only intensified the Virginia cyclone but caused a strange quirk in its course. Instead of moving northeastward along the Atlantic seaboard, the jet carried it northwestward through west central Pennsylvania and then toward the eastern end of Lake Erie. As it started on this aberrant track, the winds over Lake Huron quickly shifted to reflect the dominance of a new control.

By Sunday evening when the storm was at its peak on Lake Huron, the central pressure was 970 millibars (28.64 inches of

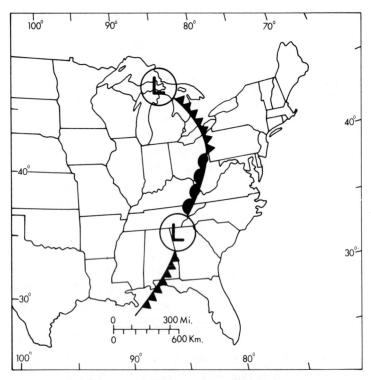

**Figure 83 Wave Development along Cold Front,
8:00 A.M. EST, 8 November 1913**

mercury). The center of the cyclone was located near Buffalo,
New York, after having moved from Virginia in only 12 hours.
Over Lake Huron, increasing north-northeasterly winds blowing along the long axis of the lake had whipped up monstrous
waves. Unfortunately anticipating the westerly winds accompanying the old cyclone over northern Lake Huron which had
long since disappeared from the weather map, the captains had
steered along the west shore of Lake Huron, attempting to
avail themselves of the protective lee of the land where little
chance existed that the west winds would drive them ashore.
The quick change of wind to the northeast caught them
unprepared and in an extremely vulnerable position.

All day Sunday the storm shrieked and howled over Lake

Huron, reaching a furious peak that evening. Then it slowly loosened its hold on the Lakes and moved northward into Canada. It brought record snowfalls to the western sections of Pennsylvania, Maryland, and West Virginia during its northwestward path from Virginia. The storm buried Cleveland under a 22-inch (56-centimeter) blanket of snow, destroyed a park project in Chicago which had taken many years to build, battered a new breakwater in Milwaukee, and downed telephone and telegraph lines throughout Michigan and Ontario. But it reserved its greatest destruction for Lake Huron.

The 1950 cousin

Nothing like the storm of 1913 in memory had happened before. But something similar to it did happen again. The storm of 25–26 November 1950 closely paralleled that of 1913, although the strongest winds in this instance occurred along the Atlantic coast. Once again, a precursor storm over the upper Great Lakes extended a front southeastward into western Pennsylvania and from there southward into northern Georgia. A wave along the front quickly intensified, and a newly formed low-pressure area moved northward through Virginia and into central Pennsylvania (figure 82). Then a new center formed over eastern Ohio and, for a short time, two low-pressure areas existed simultaneously. The Ohio center then became dominant, and it slowly completed a loop over northeastern Ohio before drifting across Lake Erie and into Ontario. Lowest pressure during its looping phase in eastern Ohio was 978 millibars (28.88 inches of mercury).

Its mode of formation and approach path to the Lakes was highly reminiscent of the storm of 1913. The author remembers the 1950 storm as a high school student in southern Michigan. Even to an inexperienced weather observer, it was readily apparent that something very unusual was going on. The north-northwest wind which arose in Michigan as the cyclone began to leave Virginia and follow its strange northwestward track was unlike any northwest wind we had ever experienced. It moaned steadily in a deep pitched voice, lacking the fitfulness which usually characterized the northwest wind in Michigan during the winter. The sky took on a dirty yellowish glow which I have not seen since.

The storm of 1950, like its 1913 relative, brought excessive snows to the upper Ohio valley. Western Pennsylvania and eastern Ohio were buried under as much as 30 inches of snow. In Columbus, Ohio, a memorable battle was waged between football teams from the University of Michigan and Ohio State University. With 13 inches (76 centimeters) of snow on the playing field, the University of Michigan managed to eke out a 3–0 victory. In Parkersburg, West Virginia, 34.4 inches (87.3 centimeters) of snow occurred, which not only exceeded the record for a single November snowfall but also the record monthly amount by 5.3 inches (13.5 centimeters). Pickens, West Virginia, recorded 57 inches (144.8 centimeters) of snow and Elkins, West Virginia, recorded 33.8 inches (85.8 centimeters).

The 1913 storm was by far much more damaging for Great Lakes shipping. While the 1950 storm caused few losses on the Lakes, it was a prolific snow bringer, and in most areas the amount of snow received exceeded that of the 1913 storm.

More important, the detailed data gathering and communication facilities of the Weather Bureau in 1950 allowed that storm to be dissected much more closely, and the resulting knowledge of its behavior permitted weathermen to learn more about its earlier predecessor. Even in 1950, however, the Weather Bureau had great difficulty predicting such a highly unusual weather feature. The value of extensive upper-air analysis for the prediction of surface weather events was just beginning to be appreciated in 1950. In 1913, of course, this was hardly even considered. In the present decade, the incorporation of upper-air data into six-level numerical models would undoubtedly give forecasters insight into the development of such a rare type of cyclone and its subsequent abnormal track. It is doubtful that such a storm could again sneak unheralded into the Lakes area without adequate warning from the Weather Service.

A legend

The great storm of 1913 has become a legend of the Lakes. A report made by the Lake Carriers Association best epitomizes this unforgettable weekend in November:

No lake master can recall in all his experience a storm of such unprecedented violence with such rapid changes in the direction of the wind and its gusts of such fearful speed. Storms ordinarily of that velocity do not last over four or five hours, but this storm raged for sixteen hours continuously at an average velocity of sixty miles per hour, with frequent spurts of seventy and over.

Obviously, with a wind of such long duration, the seas that were made were such that the lakes are not ordinarily acquainted with. The testimony of masters is that the waves were at least thirty-five feet high and followed each other in quick succession, three waves ordinarily coming right after the other.

They were considerably shorter than the waves that are formed by the ordinary gale. Being of such height and hurled with such force and rapid succession, the ships must have been subjected to incredible punishment.

Masters relate that the wind and sea were frequently in conflict, the wind blowing one way and the sea running in the opposite direction. This would indicate a storm of cyclonic character. It was unusual and unprecedented and it may be centuries before such a combination of forces may be experienced again (Ratigan 1960, pp. 113–14, reprinted by permission).

The Armistice Day Storm of 1940

Twenty-seven years later, almost to the day, another great storm on the Lakes plundered shipping and battered coastlines. The Armistice Day storm of 11 November 1940 was, in the opinion of Captain Harold B. McCool, skipper of the *Crescent City,* "even more severe than the disastrous storm during the fall of 1913" (Knarr 1951).

There were some sharp differences between the two storms, however, as well as some likenesses. The Armistice Day storm chiefly affected Lake Michigan, in contrast to the 1913 storm which, for the most part, spared that lake. The Armistice Day storm did not approach the Lakes from an unusual direction but followed one of the more common storm tracks out of Oklahoma and Kansas, through the upper peninsula of Michigan, then across Lake Superior into Canada—a track not unlike the November 1975 storm which destroyed the *Edmund Fitzgerald.* The extreme intensity of the low-pressure center placed it in a class of its own, however. The barometer reading at Houghton, Michigan, of 996 millibars (28.57 inches of mercury) when the storm center was located just to the west was one of the lowest pressures ever recorded in the Lakes area,* comparable to the pressure within the eye of a hurricane. Unlike the storm of 1913, the powerful winds accompanying the Armistice Day storm blew from the southwest. The southwest winds, the track of the storm, and the north-south alignment of Lake Michigan conspired to concentrate most destructive effects on that lake.

As in 1913, a sudden wind shift caught the vessels on the wrong side of the lake and quickly rendered them helpless, buffeted and driven onshore by 50-mile-per-hour (80-kilometer-per-hour) winds. While the ships on Lake Huron during the 1913 disaster were caught along the west shore when

*Not as low, however, as the pressures recorded during the 1978 blizzard.

the wind suddenly shifted from northwest to northeast, the Armistice Day storm began with strong southeast winds. A shift to the southwest caused the ships to flounder along the east shore of Lake Michigan, where they had sought protective shelter.

The Weather Bureau, in a tally of deaths and casualties, stated that the most severe effects in addition to the loss of ships and men on Lake Michigan occurred within the states of Illinois, Michigan, and Minnesota. Forty-nine persons lost their lives in Minnesota, and loss of livestock and poultry, damage to highways, trees, and communication facilities totaled over $1.5 million. Thirteen deaths occurred in Illinois along with over $2 million in damages. In Michigan, four deaths resulted, several million dollars worth of damage was incurred by communications facilities (including loss of the WJR Detroit radio transmitter), and in the northern part of the state, acres of forest were uprooted.

On Lake Michigan, the *Wm. B. Davoc* with a loss of thirty-three men, the *Anna C. Minch* with a loss of twenty-four, and the *Novadoc* with a loss of two—all went down near Pentwater, Michigan. Many other ships were battered and driven ashore.

So powerful were the winds that the waters of the Great Lakes were blown away from the upwind shores and piled up on the downwind sides. A drop of Lake Michigan by 4.8 feet (1.4 meters) at Chicago was compensated by a rise of 4.5 feet at Beaver Island in northern Lake Michigan. At Toledo, Ohio, the level of Lake Erie dropped to 566.3 feet (172 meters) mean sea level, while at Buffalo, the eastern extremity, the level of the water rose to 578.8 feet (176 meters) mean sea level, a range very close to the record of 14 feet (4.2 meters).

Origins of the Armistice Day storm

The storm itself began as a secondary disturbance in Idaho

on the evening of 8 November. Its existence, unlike the 1913 storm which took the Weather Bureau by surprise, was thus known for more than 2 days before it entered the Great Lakes region. From Idaho, the cyclone drifted southeastward to southern Colorado, then paused, as if deciding its future course and behavior (figure 82).

Early on the morning of 10 November, the cyclone again began to drift eastward along the border between the states of Kansas and Oklahoma. By evening it was located in south central Kansas; then things began to happen. A favorable alignment of the jet began to intensify the storm rapidly and divert it from its direct easterly heading to a bearing north by northeast. Extremely cold air from the plains of Canada began to pour southward to the rear of the developing storm and, to the east of the storm center, warm, moist air from the Gulf of Mexico streamed northward toward the Great Lakes.

The clash of unseasonably cold air with warm Gulf air released large amounts of potential energy to fuel the intensifying cyclone and maintain the powerful steering jet aloft. After moving slowly through Colorado and western Kansas, the storm quickly accelerated to 35 miles per hour as it began its track north-northeast toward the western tip of Lake Superior.

On the morning of 11 November, it was located over north central Iowa and, by noon, had moved to west central Wisconsin, with central pressures dropping to near hurricane values. At Chicago that morning, the passage of a Pacific front occurred at about 9:00 A.M., accompanied by a quick shift of the wind from southeast to southwest and a pickup in wind velocity. It was this wind shift which was to prove fatal to the Lakes ships hugging the east shore of Lake Michigan for protection from the increasing southeast winds.

A second front passed through Chicago at 11:30 A.M., ushering in colder polar continental air. Little change of wind direction occurred with the passage of this front, but it was accompanied by a sharp pickup of wind velocity. The stage was now set for one of the wildest afternoons Lake Michigan had ever experienced.

From noon until evening, wind velocities at the Chicago Weather Bureau station averaged 35 miles per hour (56 kilometers per hour) with gusts to 65 miles per hour (105 kilometers per hour) recorded between 3:00 and 4:00 P.M. As the center of the storm moved quickly from near La Crosse, Wisconsin, to just west of Houghton, Michigan, where it was located at 6:30 P.M., wind gusts of 80 miles per hour (129 kilometers per hour) at Grand Rapids, Michigan, 67 miles per hour (108 kilometers per hour) at Muskegon, Michigan, and 61 miles per hour (98 kilometers per hour) at Alpena, Michigan, were recorded. These were the highest winds of record at these locations. Lake Michigan became a turmoil, with the southwest winds developing maximum fetch along the long axis of the lake. Because Armistice Day was a holiday and fell on Monday in 1940, many persons were caught returning after long weekends away from home, and hunters and outdoorsmen were unable to find shelter. To the west of the storm track over western Iowa and much of Minnesota, heavy snow fell, up to 22 inches (55.8 centimeters) in parts of Minnesota.

The barometer reading at Houghton, Michigan (966 millibars or 28.57 inches of mercury), was the lowest pressure recorded within the United States during the passage of the storm, although the pressure continued to fall as the storm crossed Lake Superior and headed into Ontario. The track of the cyclone center closely resembled that of the 1975 storm which sank the *Fitzgerald*, although the 1940 center moved about 100 miles (161 kilometers) west of the 1975 cyclone. In terms of intensity and wind velocities, however, it exceeded the more recent storm.

The 1913 and 1940 storms are the worst in memory. Each year has its quota, and there will be other storms just as severe but, one hopes, less damaging. Stark testimony to the continued vulnerability of men and vessels on the Great Lakes is provided by the photographs of the torn remnants of the *Edmund Fitzgerald* and the haunting refrain of a popular song by which the ship is memorialized.

The Future

THE CONCLUSION IS INESCAPABLE THAT WEATHER
and climate will become increasingly important to future
generations of Great Lakes residents. The *Fitzgerald* disaster
demonstrated that Great Lakes shipping has not achieved
immunity to weather hazards. Current efforts to extend the
shipping season through the winter months will add to weather-
related risks incurred by the lake vessels. Better prediction of
storms, increased knowledge of ice cover, possible regulation
of channel depths through cloud seeding—all may be desirable
to insure the safety of lake commerce.

Changes of climate may force different approaches toward
maintaining agricultural productivity. The likelihood of increased weather variability during future decades places agriculture within a framework of climatic insecurity not previously experienced in the twentieth century. Weather modification may become a standard agricultural practice—no longer a
luxury, but a necessity designed to dampen the effect of
increasingly erratic weather patterns.

Recreation and tourism, of major importance in the Great
Lakes area, are demonstrably weather-dependent. Long-term
climatic changes may be either beneficial or detrimental to these

305

industries. The role of weather modification as a practical appendage to the recreation industry will come under much closer scrutiny. The severity of shoreline erosion during the early 1970s also promises to place the value of weather modification in sharper focus for control of lake levels. Can precipitation redistribution or suppression be useful for alleviating shoreline erosion by reducing the level of the Lakes during high-water periods?

As the cities continue to grow, the interactions between urban effects and lake effects will become more acute. Can effective urban and regional planning be employed to forestall the occurrence of inadvertent weather effects which may prove undesirable? Which of the possible weather effects *are* undesirable?

As these questions are posed in the future, it is certain that answers will not come easily. Increased knowledge regarding the behavior of the Lakes and the atmosphere which overlies them is greatly needed. Fortunately, numerous scientific groups have been involved in such research. No international boundaries or customs regulations restrict the flow of scientific information about the Lakes. Shared by two nations, research on the Lakes has been truly international. National agencies of both Canada and the United States have worked jointly. Among these in Canada have been the Department of Energy, Mines and Resources with its Great Lakes Research Division; the Department of the Environment with its Atmospheric Environment Service and Canada Centre for Inland Waters; and the Ministry of Transport and the National Research Council. In the United States, the National Oceanic and Atmospheric Administration, the Army Corps of Engineers, the Environmental Protection Agency, and the National Aeronautics and Space Administration, among others, have been actively engaged in Great Lakes Research. Many provincial and state agencies, particularly the Illinois State Water Survey and the Ontario Ministry of the Environment—and university groups such as the Great Lakes Research Division of the University of Michigan, the Great Lakes Center of the University of Wisconsin-Milwaukee, and the Great Lakes Institute of the University of Toronto—have also been involved.

International Field Year for the Great Lakes

Of particular interest in the acquisition of knowledge regarding Great Lakes weather have been several large-scale coordinated research efforts involving a consortium of agencies and individuals. The *International Field Year for the Great Lakes* (*IFYGL*) was a joint effort by the United States and Canada to learn more about the Great Lakes (Ludwigson 1974). It was a part of the International Hydrological Decade initiated by UNESCO (United Nations Educational and Scientific Organization). The decade began on 1 January 1965 and ended on 31 December 1974.

IFYGL designated Lake Ontario as the site of a concentrated 12-month research program by public and private agencies of the United States and Canada. An intensive data-gathering effort began in April 1972. The data-gathering equipment included five research vessels, a network of instrumented buoys, a number of radars for the estimation of precipitation over the lake, a precipitation network, several instrumented aircraft, six rawinsondes for collecting upper-air data, a number of shoreline meteorological stations, and the use of satellites and their imagery.

A number of scientific programs were established under the IFYGL umbrella. A program to learn more about the flow of moisture into and from Lake Ontario (*Terrestrial Water Balance Program*) attempted to make more precise measurements of the water inflow and outflow and of precipitation and evaporation over the lake. A program entitled *Lake Meteorology and Evaporation* directed its efforts into learning more about the atmospheric water balance over the lake as a function of altitude and time and also attempted to make precise estimates of evaporation and precipitation within the lake basin.

Since the heat stored within the lake plays a basic role in the instigation of lake weather effects and is also important to shipping and agricultural interests, a *Lake Energy Balance* program attempted to quantify the energy flow to and from the lake. Studies were made of solar radiation over the lake, the lake's albedo, and the role of ice and snow in altering energy transfer. Longwave radiation to and from the lake surface, and sensible and latent heat transfer were also studied.

The circulation within the lake, including wave mechanisms, was studied by a *Water Movements* program, and an *Atmospheric Boundary Layer* program measured the flow of moisture and heat from the surface of the lake into the atmosphere.

An immense storehouse of data was gathered to answer questions about the behavior of the Lakes, both for present-day concerns and also in anticipation of problems which might arise in the future. The data from IFYGL became public property for release to a variety of users, including scientific personnel, water-quality managers, and those involved with operations possibly sensitive to the environment, such as shipping, air and ground transport, nuclear plants, and so on. The data archives for Canada are located at the Canada Centre for Inland Waters, and those of the United States are at the National Climatic Center, Asheville, North Carolina.

The Chicago Area Program

The southern Lake Michigan area is of very special interest for meteorologists and climatologists. Here the lake effects combine with a variety of potential inadvertent modifications. It is in this area also that deliberate modification may become a viable weather control option in the future. The shores of southern Lake Michigan are rimmed with industries, highways, and people. It seems only natural that a concentrated effort by

**The radar utilized in the Chicago Area Program to gather data on
snowstorms and cloud physics from a site
in Muskegon, Michigan.**

scientists should be made to gain an understanding of the
weather and climate of this region.

The *Chicago Area Program (CAP)* (Changnon and Semonin 1978) is an extension of the METROMEX instrumental
data-gathering system into the southern Lake Michigan area.
An area of over 20,000 square miles, including southern Lake
Michigan, Wisconsin, Illinois, Indiana and Michigan, has been
chosen for a four- to six-year study of precipitation characteristics (Changnon and Huff 1976). More than three hundred
recording rain gauges with hail sensors (small styrofoam pads
with thin sheets of aluminum foil covering them that will record
the indentations of hailstones) have been placed along the
shores of the lake. Precipitation over the lake will be measured
by two special weather radars with calibration techniques based

upon readings of the lakeside recording gauges. Operation of the gauges began in the summer of 1976.

There are a number of reasons why the southern Lake Michigan region was selected for study as a logical extension of the comprehensive METROMEX project. The population and size of Chicago have an order of magnitude larger than St. Louis. Whereas St. Louis contains 1.5 million persons in a 1,900-square-mile area, Chicago contains 7.4 million people in a 3,600-square-mile area. Thus the processes identified as occurring at St. Louis can be investigated as they occur in an even larger urban area.

Chicago is located on the shore of Lake Michigan, and an intermixture of urban and marine effects may be expected to complicate its weather and climate. A look at a map shows that many of the largest cities of the world are located in coastal zones where similar complications from urban and marine effects may occur. It is desirable to develop techniques for recognizing the individual impact of each effect and of separating the urban effects from the marine effects. This is an immense and complicated but necessary task. The Chicago area, where considerable previous work has been done, would seem a natural selection for this challenging job.

Then there is the suspected presence of one of the largest precipitation anomalies in the world, at La Porte, Indiana. As mentioned previously, a 30% rainfall increase has occurred over the years. This has been attributed to the effect of upwind industrialization. In light of the more recent METROMEX findings, it is not surprising that such an effect exists, but the dimensions of the anomaly still boggle the minds of scientists. Although the controversy has simmered in recent years, it has never been completely resolved. It is to be hoped that the new data from the Chicago Area Project will settle the La Porte question once and for all.

The specific goals of the program include, first, the study of inadvertent precipitation modification. The large-sized anomaly downwind from Chicago suggests that the Chicago metropolitan area, with its large population and concentrations of industry and particularly with its steel-producing industries (which supply ice-forming nuclei to the air), may be capable of significantly changing the precipitation patterns within the region. Chicago is a snowier area than St. Louis, and an appraisal of the city's role in affecting snowfall will be a secondary goal of the inadvertent precipitation study. Lake-effect snowstorms which form over southern Lake Michigan will also be carefully examined in order to determine whether the heat and pollution of the Chicago area play a role in instigating and maintaining these systems.

Another specific goal is to investigate the role of Lake Michigan in altering precipitation patterns over and around the lake. Past studies have had to rely on data from the National Weather Service installations at scattered locations along the Lakeshores. These installations were usually nonrecording gauges. Thus little could be learned concerning the ways in which the lake altered intensities and hourly frequencies of precipitation. Few data were available to describe how much (or how little) precipitation occurs over the lake. The recording gauge network of CAP plus the scanning radars will allow scientists to learn much more than the details of how the lake alters precipitation patterns.

A third goal focuses on the frequencies and intensities of rainfall within the Chicago urban area itself. A sophisticated but man-controlled hydrologic system governs the flow of water in the Chicago urban area between Lake Michigan and the Illinois River drainage network. The system is highly sensitive to large inflows of water in the form of intense rainshowers. Better information regarding the likelihood of intense rainshower occurrence over the Chicago urban area will allow more finely tuned decision making by engineers and technicians. In terms of planning, additional rainfall information over the urban area will be valuable for engineers, for the design of sewer districts, for hydrologic consultants, and for a large range of metropolitan planning activities. For predictive purposes, the information supplied by the two weather radar installations associated with the project will be of immense value.

Since the southern Lake Michigan shoreline is ringed with a number of power installations, both thermal and nuclear, a fourth goal of CAP will be to study the effects of these plants on precipitation. Is there downwind enhancement from power plant activities? And if so, what are the possible implications for the future as energy needs continue to grow?

A fifth goal of CAP is to study the transport and dispersal of pollutants by means of an atmospheric chemistry program. The southern Lake Michigan area, as mentioned, poses some very special problems involving atmospheric pollution. Further studies are necessary in order to formulate the air control principles and planning measures for the future.

In all of these lines of research, the technical knowledge and experience gained from the 5-year METROMEX project will be of an immense value. The Illinois State Water Survey, which has long been involved in studying the surface and atmospheric water resources of the area, has been the primary agency involved in the CAP program. Other interested groups will also participate. Like the IFYGL project, the data are to be made available to a wide variety of users.

Research programs such as IFYGL and CAP should unravel some of the unknowns about those unique weather factories called the Great Lakes. But many mysteries will remain. The Lakes will continue to taunt meteorologists with their unique brands of weather, the intricacies of which may never be completely understood. In spite of efforts to predict, tame, and alter, Great Lakes weather can be counted on to generate the unforeseen and unexpected—to offer a varied program of weather from a vast repertoire of meteorological surprises.

Capricious, occasionally dangerous, but often benign, the kaleidoscopic weather of the Great Lakes is never dull. To understand it is a challenging goal.

References

Barcus, Frank A. 1960. *Freshwater fury*. Detroit: Wayne State University Press.

Bellaire, Frank R. 1965. The modification of warm air moving over cold water. *Proceedings of the Eighth Conference on Great Lakes Research*, pp. 249–56. Ann Arbor: University of Michigan.

Brooks, C. F. 1914. The snowfall of the eastern United States. *Monthly Weather Review* 43 (Jan. 1914): 2–11.

Bryson, Reid A. 1966. Airmasses, streamlines and the boreal forest. *Geographical Bulletin* 8, no. 3 (1966): 2283—69.

———. 1974. A perspective on climatic change. *Science* 184, no. 4138: 753–60.

Bryson, Reid A., and Baerreis, David A. 1967. Possibilities of major climatic modification and their implications: Northwest India, a case study. *Bulletin American Meteorological Society* 48, no. 3 (March 1967): 136–42.

Byers, Horace R. 1974. *General Meteorology*. New York: McGraw-Hill.

Changnon, Stanley A., Jr. 1957. *Thunderstorm-precipitation relations in Illinois*. Report of Investigation 34. Urbana: Illinois State Water Survey.

———. 1968a. The La Porte weather anomaly: Fact or fiction. *Bulletin American Meteorological Society* 49, no. 1 (Jan. 1968): 4–11.

314   References

————. 1968*b. Precipitation climatology of Lake Michigan basin.* Urbana: Illinois State Water Survey.

Changnon, Stanley A., Jr., and Huff, F. A. 1976. A multi-purpose hydrometeorological system for urban hydrology applications. *Preprints, Conference on Hydro-Meteorology,* pp. 42–47. Boston: American Meteorological Society.

Changnon, Stanley A., Jr., and Jones, Douglas M. A. 1972. Review of the influences of the Great Lakes on weather. *Water Resources Research* 8, no. 2 (April 1972): 360–71.

Changnon, Stanley A., Jr., et al. 1977. *Hail suppression impacts and issues.* Urbana: Illinois State Water Survey.

Changnon, Stanley A., Jr., and Semonin, Richard G. 1978. Chicago Area Program: A major new atmospheric effort. *Bulletin American Meteorological Society* 59, no. 2 (Feb. 1978): 153–60.

Cole, Alan L. 1971. Hindcast waves for the western Great Lakes. *Proceedings of the Fourteenth Conference on Great Lakes Research,* pp. 412–21. Ann Arbor: International Association for Great Lakes Research.

Cole, Henry S., and Lyons, Walter A. 1972. The impact of the Great Lakes on the air quality of urban shoreline areas: Some practical applications with regard to air pollution control policy and environmental decision making. *Proceedings of Fifteenth Conference on Great Lakes Research,* pp. 436–63. Ann Arbor: International Association for Great Lakes Research.

Conrad, Victor, and Pollak, L. W. 1950. *Methods of Climatology.* Cambridge, Mass.: Harvard University Press.

Eichenlaub, Val L., and Bacon, Christina M. 1974. Convective precipitation in the Detroit urban area. *The Professional Geographer* 27, no. 2 (May 1974): 140–46.

Eichmeier, A. W. 1951. Snowfalls—Paul Bunyan style. *Weatherwise* 4 (Sept. 1951): 124–27.

Huff, F. A.; Beebe, R. C.; Jones, D. M. A.; Morgan, G. M., Jr.; and Semonin, R. G. 1971. *Effect of cooling tower effluents on atmospheric conditions in northeastern Illinois.* Circular 100. Urbana: Illinois State Water Survey.

Huff, F. A., and Changnon, Stanley A., Jr. 1972. *Climatological assessment of urban effects of precipitation*. Urbana: Illinois State Water Survey.

Jiusto, James E., and Holroyd, Edmond W., III. 1970. *Great Lakes Snowstorms*, part 1, *Cloud Physics Aspects*. Albany: State University of New York, Atmospheric Sciences Research Center.

Jiusto, James E.; Paine, Douglas A.; and Kaplan, Michael L. 1970. *Great Lakes Snowstorms*, part 2, *Synoptic and Climatological Aspects*. Albany: State University of New York, Atmospheric Sciences Research Center.

Knarr, Aural J. 1951. The Midwest storm of November 11, 1940. *Monthly Weather Review* 69, no. 6 (June 1951): 169-78.

Kopec, Richard J. 1967. Areal patterns of seasonal temperature anomalies in the vicinity of the Great Lakes. *Bulletin American Meteorological Society* 48, no. 12: 884-89.

Lahey, J. F.; Bryson, R. A.; Wahl, E. W.; and Henderson, V. D. 1950. *Atlas of 500 mb wind characteristics for the Northern Hemisphere*. Madison: University of Wisconsin Press.

Lamb, H. H. 1966. Climate in the 1960's. *Geographical Journal* 132 (1966): 183-212.

Lavoie, R. L.; Cotton, W. R.; and Hovermale, J. B. 1970. *Investigations of lake effect storms*. State College, Pa.: Pennsylvania State University, Department of Meteorology.

Ludwigson, John D. 1974. *Two nations, one lake—science in support of Great Lakes management: Objectives and activities of the International Field Year for the Great Lakes 1965-1973*. Ottawa: Environment Canada.

Lyons, Walter A. 1966. Some effects of Lake Michigan upon squall lines and summertime convection. *Proceedings Ninth Conference on Great Lakes Research*, pp. 259-73. Ann Arbor: University of Michigan.

Lyons, Walter A., and Wilson, John W. 1968. *Control of summertime cumuli and thunderstorms by Lake Michigan during non-lake breeze conditions*. Department of Geophysical Sciences, Satellite and Mesometeorology Research Project Research Paper No. 74. Chicago: University of Chicago.

Lyons, Walter A., and Pease, Steven R. 1972. Steam devils over Lake Michigan. *Monthly Weather Review* 100, no. 3 (March 1972): 235–37.

Mitchell, C. L. 1921. Snow flurries along the eastern shore of Lake Michigan. *Monthly Weather Review* 49 (Sept. 1921): 502–3.

Muller, Robert A. 1966. Snowbelts of the Great Lakes. *Weatherwise* 19, no. 6 (Dec. 1966): 248–55.

Nurnberger, Fred V. 1974. Review of weather modification legislation. Unpublished paper.

Petterssen, Sverre. 1969. *Introduction to Meteorology*. 3d ed. New York: McGraw-Hill.

Ratigan, William. 1960. *Great Lakes shipwrecks and survivals*. Grand Rapids: Eerdmans.

Rockwell, David C. 1966. Theoretical free oscillations of the Great Lakes. *Proceedings Ninth Conference on Great Lakes Research,* pp. 352–68. Ann Arbor: University of Michigan.

Ryznar, Edward, and Weber, Michael R. 1976. *An investigation of the meteorological impact of mechanical-draft cooling towers at the Palisades nuclear plant*. Ann Arbor: University of Michigan, Department of Atmospheric and Oceanic Science.

Saunders, Peter M. 1964. Sea smoke and steam fog. *Quarterly Journal of the Royal Meteorological Society* 90, no. 384 (April 1964): 155–65.

Schaefer, Vincent. 1968. The early history of weather modification. *Bulletin American Meteorological Society* 49, no. 4 (April 1968): 337–42.

Smith, Allen R. 1966. An investigation of the surface temperature distribution in the Detroit region. Master's Thesis, Department of Geography. Kalamazoo, Mich.: Western Michigan University.

Smith, R. 1971. *The climate of Toronto*. Toronto: Canada Department of Environment, Atmospheric Environment Service.

Stout, Glenn E., and Ackermann, William C. 1974. *Hydrologic-economic feasibility study of precipitation augmentation over the Great Lakes*. Urbana: Illinois State Water Survey.

Strahler, Arthur N. 1961. *Physical Geography* 2d ed. New York: Wiley and Sons.

Strommen, Norton D. 1975. Seasonal change in the axis of maximum lakesnow in western lower Michigan. Ph.D. dissertation, Department of Geography, Michigan State University.

Strong, Alan E., and Bellaire, Frank R. 1965. The effect of air stability on wind and waves. Proceedings Eight Conference on Great Lakes Research, pp. 294–310. Ann Arbor: University of Michigan.

Terjung, Werner H. 1966. Physiologic climates of the coterminous United States: A bioclimatic classification based on man. *Annals Association of American Geographers* 56, no. 7 (March 1966): 141–79.

United States Environmental Data Service. 1968. *Climatic atlas of the United States*. Washington, D.C.: U.S. Government Printing Office.

Webb, T., and Bryson, R. A. 1972. Lake and post-glacial climatic change in the northern Midwest: Quantitative estimates derived from fossil pollen spectra and multivariate statistical analysis. *Quaternary Research* 2 (1972): 70–115.

Weickmann, Helmut. 1972. Man-made weather patterns in the Great Lakes basin. *Weatherwise* 25, no. 6 (Dec. 1972): 260–67.

Wiggin, B. L. 1950. Great snows of the Great Lakes. *Weatherwise* 3, no. 6 (Dec. 1950): 123–26.

Appendix:
Some Climatic Extremes within the Great Lakes Basin

Maximum Temperature
  Mio, Mich.   112°F (44.4°C)   July 13, 1936
  Fond du Lac, Wis.   110°F (43.3°C)   [Date unknown]

Minimum Temperature
  White River, Ont.   −61°F (−51 6°C) Jan. 1888
  Danbury, Wis.   −54°F (−47.7°C) Jan. 24, 1922
  Stillwater Reservoir,
    N.Y.   −52°F (−46.6°C) Feb. 9, 1934
  Vanderbilt, Mich.   −51°F (−46.1°C) Feb. 9, 1934

Most Rainfall, 24 Hours	*In.*	*Mm.*	
Mellen, Wis.	11.72	(297)	June 24, 1946
Mahnomen, Minn.	10.75	(273)	June 20, 1909
Sandusky, Ohio	10.51	(267)	June 12, 1966
Erie, Pa.	10.42	(264)	June 22, 1947
Bloomingdale, Mich.	9.78	(248)	Sept. 1, 1914
Snelgrove, Ont.	7.15	(181)	Oct. 15, 1954

NOTE: Values given are unofficial and pertain to the drainage basin of the Great Lakes only.

319

Most Rainfall, Month

	In.	Mm.	
Battle Creek, Mich.	16.24	(412)	June 1883
Kitchener, Ont.	13.90	(353)	July 1915

Most Rainfall, Year

	In.	Mm.	
Adrian, Mich.	64.01	(1625)	1881
Stratford, Ont.	63.79	(1620)	1884
Embarrass, Wis.	62.07	(1576)	1884

Least Rainfall, Year

	In.	Mm.	
Oshawa, Ont.	7.76	(197)	1919
Plum Island, Wis.	12.00	(304)	1937
Croswell, Mich.	15.64	(397)	1936
Lewiston, N.Y.	17.41	(442)	1941

Most Snowfall, 24 Hrs.

	In.	Cm.	
Adams, N.Y.	68	(172)	Jan. 9, 1976
Barnes Corners, N.Y.	54	(137)	Jan. 9, 1976
Watertown, N.Y.	45	(114)	Nov. 14, 1900
Pelee Island, Ont.*	36	(91)	Feb. 1905
Ishpeming, Mich.	29	(74)	Feb. 23, 1922
Pigeon River Bridge, Minn.	28	(71)	Apr. 4, 1933

NOTE: Values given are unofficial and pertain to the drainage basin of the Great Lakes only.

*Other New York snowbelt stations have exceeded this value.

Most Snowfall, One storm

	In.	Cm.	
Oswego, N.Y.	102	(259)	Jan. 27–31, 1966
Watertown, N.Y.	69	(175)	Jan. 18–22, 1940
Adams, N.Y.	68	(173)	Jan. 9, 1976
Oswego, N.Y.	66.7	(169)	Dec. 7–11, 1958
Barnes Corner, N.Y.	54	(137)	Jan. 9, 1976
Boonville, N.Y.	52	(132)	Nov. 22–23, 1976
Calumet, Mich.**	46	(117)	Jan. 15–20, 1950
La Porte, Ind.	37	(95)	Feb. 14–19, 1958

Most Snowfall, Month

	In.	Cm.	
Hooker, N.Y.	149	(378)	Jan. 1977
Hooker, N.Y.	137	(348)	Jan. 1976
Old Forge, N.Y.	120	(305)	Mar. 1971
Calumet, Mich.	115.3	(292)	Jan. 1950
Rocklyn, Ont.	107.5	(373)	Dec. 1903

Most Snowfall, Season

	In.	Cm.	
Hooker, N.Y.	466.9	(1186)	1976–77
Old Forge, N.Y.	408.3	(1037)	1976–77
Bennett Bridge, N.Y.	388.5	(987)	1976–77
Old Forge, N.Y.	375	(952)	1970–71
Barnes Corner, N.Y.	370	(940)	1976–77
Bennett Bridge, N.Y.	351.9	(894)	1938–39
Tahquamenon Falls, Mich.	332.8	(846)	1976–77
Herman, Mich.	308.4	(782)	1975–76
Steep Hill Falls, Ont.	301.5	(764)	1938–39

NOTE: Values given are unofficial and pertain to the drainage basin of the Great Lakes only.

**Exceeded by other stations within New York State snowbelts.

Acknowledgments

The authors and publishers cited below have generously permitted the use of their work in the illustrative material for this book:

Figure 2. Redrawn from International Great Lakes Levels Board, *Regulation of Great Lakes Water Levels*, p. 23 fig. 4. Washington, D.C.; Ottawa: International Joint Commission, 1973.

Figure 3. Reprinted with permission from Morris Neiburger, James G. Edinger, and William D. Bonner, *Understanding Our Atmospheric Environment*, fig. 3.2, p. 43. W. H. Freeman and Company. Copyright © 1973.

Figure 16. Redrawn from Glenn T. Trewartha, *An Introduction to Climate*, fig. 6.23, p. 219. New York: McGraw-Hill, 1968.

Figures 20 and 21. Redrawn from Arnold Court, "The Climate of the Coterminous United States" in Helmut Landsberg, R. A. Bryson, and F. K. Cole, eds., *Climates of North America: World Survey of Climatology II,* fig. 23, p. 222; fig. 28, p. 230. Amsterdam: Elsevier, 1973.

Figure 26. Redrawn from G. K. Rodgers, "The Thermal Bar in Lake Ontario Spring 1965 and Winter 1965–66" in *Proceedings Ninth Conference on Great Lakes Research*, fig. 3, p. 372. Ann Arbor: University of Michigan, 1966.

Figure 27. Data from Phil E. Church, *The Annual Temperature Cycle of Lake Michigan,* 30; part 1, *Cooling from Late Autumn to the*

Terminal Point, 1941 –42; part 2, *Spring Warming and Summer Stationary Periods, 1942.* Chicago: University of Chicago, Department of Meteorology, 1942.

Figure 28. Redrawn from Frederick G. Millar, "Surface Temperatures of the Great Lakes" in *Journal of Fisheries Research Board of Canada* 9 (1952): 329–76.

Figure 29. Redrawn from maps in Donald R. Rondy, *Great Lakes Ice Atlas,* NOAA Technical Memorandum No. 5 LSR 1. Washington, D.C.: Department of Commerce, 1971.

Figure 37. Redrawn from Walter A. Lyons, "Some Effects of Lake Michigan upon Squall Lines and Summertime Convection" in *Proceedings Ninth Conference on Great Lakes Research,* fig. 3B, p. 262. Ann Arbor: University of Michigan, 1966.

Figure 38. Redrawn from Walter A. Lyons and John F. Chandik, "Thunderstorms and the Lake Breeze Front" in *Proceedings Seventh Conference on Severe Local Storms,* fig. 4, p. 291. Kansas City, Mo.: American Meteorological Society, 1971.

Figure 39. Redrawn from Robert J. Ressel, "What Is a Lake Breeze and Its Possible Effect on the Diurnal Temperature Curve?", fig. 2.0, p. 7. Research paper, Department of Geography. Kalamazoo, Mich.: Western Michigan University, 1970.

Figure 46. From Val. L. Eichenlaub and Elizabeth Garrett, "Climatic Modification and Lake-Effect Snowfall along the Lee Shore of Lake Michigan: A Classic Month as Viewed by Radar and Weather Satellite" in *Weatherwise* 25, no. 6 (Dec. 1972): fig. 5, p. 272.

Figure 47. Redrawn from Robert Muller, "Snowbelts of the Great Lakes" in *Weatherwise* 19, no. 6 (Dec. 1966): 250–51.

Figures 48 and 50. Redrawn and modified from Val L. Eichenlaub, "Lake Effect Snowfall to the Great Lakes: Its Role in Michigan" in *Bulletin American Meteorological Society* 51, no. 5 (May 1970): fig. 2, p. 404; fig. 8, p. 408.

Figure 51. Redrawn from Arthur N. Strahler, *Physical Geography* 2d ed., fig. 9.2, p. 142. New York: Wiley, 1961.

Figure 54. Data from Donald G. Baker and John C. Klink, *Solar Radiation, Probabilities, and Areal Distribution in the North*

Central Region, North Central Regional Research Publication No. 225. Minneapolis: University of Minnesota, 1975.

Figure 55. Data from William H. Klein, *Principal Tracks and Mean Frequencies of Cyclones and Anticyclones in the Northern Hemisphere,* Research Paper No. 40. Washington, D.C.: U.S. Weather Bureau, 1957.

Figure 56. Reprinted with the permission of the author and the American Meteorological Society from Richard J. Kopec, "Continentality around the Great Lakes" in *Bulletin American Meteorological Society* 46, no. 2 (Feb. 1965): fig. 1, p. 55.

Figures 57, 60, 61, and 62. Redrawn from D. W. Phillips and J. A. W. McCulloch, *The Climate of the Great Lakes Basin,* Climatological Studies No. 20, charts 19, 13, 11, and 21. Toronto: Environment Canada, Atmospheric Environment Service, 1972.

Figure 63. Data from *Monthly Precipitation Probabilities by Climatic Divisions,* Misc. Pub. No. 1160, pp. 75 and 96. Washington, D.C.: Departments of Commerce and Agriculture, 1969.

Figures 65 and 66. Redrawn and modified from Werner H. Terjung, "Physiologic Climates of the Coterminous United States: A Bioclimatic Classification Based on Man" in *Annals Association of American Geographers* 56, no. 1 (March 1966): fig. 8, p. 162; fig. 4, p. 154.

Figures 67, 68, 69, 70. Redrawn from Robert L. Janiskee, "Comfort Climates at Grand Rapids, Michigan: A Dynamic Approach," fig. 3, p. 35; fig. 4, p. 59; fig. 5, p. 71; fig. 6, p. 92. Master's Thesis, Department of Geography, Kalamazoo: Western Michigan University, 1969.

Figures 72 and 73. Redrawn from Val L. Eichenlaub, "Climatic Change in the Southern Great Lakes–Eastern Corn Belt," in *East Lakes Geographer* 11 (May 1976): 130, 131.

Figure 74. Redrawn from Paul F. Krause, "An Investigation of Possible Climatic Trends in the Northwestern Portion of the Great Lakes Basin," fig. 8, p. 27. Research paper, Department of Geography, Kalamazoo, Mich.: Western Michigan University, 1976.

Figure 75. Redrawn and modified from Stanley A. Changnon, Jr., "The LaPorte Weather Anomaly: Fact or Fiction" in *Bulletin American Meteorological Society* 49, no. 1 (Jan. 1968): fig. 2, p. 5.

Figures 78 and 79. Redrawn from Allen R. Smith, "An Investigation of the Surface Temperature Distribution in the Detroit Region," fig. 3, p. 72; map 8, p. 81. Master's Thesis, Department of Geography. Kalamazoo, Mich.: Western Michigan University, 1966.

Figure 80. Redrawn from Henry S. Cole and Walter A. Lyons, "The Impact of the Great Lakes on the Air Quality of Urban Shoreline Areas: Some Practical Applications with regard to Air Pollution Control Policy and Environmental Decision Making" in *Proceedings Fifteenth Conference on Great Lakes Research*, fig. 8, p. 441. Ann Arbor: International Association for Great Lakes Research, 1972.

Figure 82. Redrawn from Charles L. Bristor, "The Great Storm of November, 1950" in *Weatherwise* 4, no. 1 (Feb. 1951): fig. 1b, p. 11; from Aural J. Knarr "The Midwest Storm of November 11, 1940" in *Monthly Weather Review* 69, no. 6 (June 1941): fig. 1, p. 169; and from C. P. Mook, "The Famous Storm of November 1913" in *Weatherwise* 2, no. 6 (Dec. 1949): figs. 1 and 2, pp. 126, 127.

Figure 83. Redrawn and modified from C. P. Mook, "The Famous Storm of November 1913" in *Weatherwise* 2, no. 6 (Dec. 1949): fig. 1, p. 126.

The photographs in this book have been supplied by:
Consumers Power Company, page 243
Val Eichenlaub, pages 100, 166, 167
Jay Harman, page 139
Illinois State Water Survey, page 309
Helen Johnson, page 173
Walter Jones, page 72
Margaret Lemone, pages 124, 146
Walter A. Lyons, pages 105, 110, 111, 125, 129, 136, 251, 265

National Oceanic and Atmospheric Administration, pages 15, 38, 46, 54, 67, 76, 77, 101, 143 (photo by Ken Dewey), 159, 191 (photo by Ken Dewey), 229 (photo by Ken Dewey)
Ray Peterson, page 161
Vincent J. Schaefer, page 35
Toledo Edison Power Company, page 259
United States Coast Guard, pages 79, 85, 88
U.S. Department of Agriculture, page 185
U.S. Department of the Interior, pages 157, 250

Index